Cleaner, Greener Labs for Analytical Chemistry

&

Instrumental Analysis

By Catherine Haustein, PhD

Central College, Pella, Iowa 50219

2021

A Greener Lab Manual

Sustainability means meeting the needs of the present without compromising the ability of future generations to meet theirs, a goal that includes the interconnected dimensions: social justice, economic prosperity, and environmental integrity. Central College has a strong tradition of commitment to sustainability, from our LEED buildings and our global sustainability academic core requirement to our community policies of justice, inclusion, and non-discrimination. The college has publicly committed to be a sustainability leader in higher education, signing the Talloires Declaration in 2006 and the American Colleges and University Presidents Climate Commitment in 2007. From Sustainability Education 5-Year Strategic Plan
May 2015

Sustainability involves making choices that make the world more livable for all. It considers positive changes that protect the future. It is by nature, unselfish, responsible, and optimistic.

Sustainability compliments the college's learning goals.

Central College Integrated Learning Goals for a 2nd year student: *Each student has implemented strategies for promoting health and reducing risk in his/her life. Each student will understand that his/her choices impact local and global sustainability*

Being sustainable reduces risk, promotes health, and reduces negative local and global impact!

No lab is totally green. All use materials and generate waste. Even this lab manual has impacted the environment. We will be using deionized (DI) water and this process uses energy. This manual has focused on using less materials and safer materials. However, every lab makes an impact on the environment. This is why some labs will be done as group or partner labs—to cut down on materials and energy use. Unless otherwise directed, each student should turn in an independent lab report that avoids plagiarism.

Table of Contents

12 Principles of Green Chemistry	page 4
Responsible Use of Fume Hoods	page 5
Washing glassware	page 6
Chemical Safety	page 6
Categories of Chemical Reagents	
Rules for Handling Reagents	page 7
Central College Vermeer Science Center Laboratory Rules and Policies	page 8
The Laboratory Notebook	page 9
Use of the Balance	page 10
Experiment 1: Lab check in, balances and desiccators	page 11
Experiment 2a: Observations of a Candle Burning	page 12
Experiment 2b: Sampling	page 13
Experiment 3: Using a Spreadsheet, Buret, and Pipet	page 14
Experiment 4: Gravimetric Determination of Phosphorus in Plant Food Using Household Chemicals	page 16
Experiment 5: Titration of KHP	page 19
Experiment 6: Titration of Vinegar, Wine, and Soft drinks	page 22
Experiment 7: Acid-base Titrations with plotting & unknown acid	page 25
Experiment 8: Acid-base Titration plots using known acids	page 27
Experiment 9: Natural Indicators	page 28
Experiment 10: Buffering Action of Hair and Skin Products	page 30
Experiment 11: Buffer Action	page 32
Experiment 12: Determination of Sodium by Flame Emission	page 35
Experiment 13: Determination of Dyes in Kool-Aid Using Visible Spectroscopy	page 37
Experiment 14: Two Colorful Plant Molecules--Determination of Beta-carotene Using Visible Spectroscopy/ Identification of Betain by Fluorescence	page 39
Experiment 15: Determination of Henna in a Hair Product Using Visible Spectroscopy	page 42
Experiment 16: Determination of Riboflavin (Vitamin B_2) by Fluorescence	page 44
Experiment 17: Determination of Nitrates NO_3^-: UV screening and ion selective electrode	page 45
Experiment 18: Gas Chromatographic determination of alcohol in orange extract	page 47
Experiment 19: Gas Chromatographic determination of Methyl salicylate in Oil of Wintergreen and/or Limonene in lemon oil by Gas Chromatography	page 48
Experiment 20: Determination of L-dopa in *Mucuna pruriens* (Velvet beans) by HPLC	page 49
Experiment 21: Determination of the slope of a pH electrode	page 51
Experiment 22: Determination of Quinine by Fluorescence	page 52
Experiment 23: Determination of Iron in Wine and Beer or Pomegranate Juice by Atomic Absorption	page 54

Experiment 24: Determination of p-coumaric acid in sage using
    fluorescence and standard addition (Under Development)   page 55
Experiment 25: Identification of Polymers using Infrared Spectroscopy   page 56
Experiment 26: Determination of 4-acetamidophenol in
    analgesics using Cyclic Voltammetry   page 57
Experiment 27: Determination of Chloride by Ion Selective
    Electrode   page 58
Experiment 28 Fluorescein in Antifreeze   page 59
Experiment 29: Determination of Aspirin by HPLC   page 61
Experiment 30: Determination of Tryptophan by Fluorescence   page 62
Experiment 31: Field Lab--Carbon Dioxide Control Chart   page 64
Appendix: Common Suffixes   page 66
    Drawer Contents   page 67

12 Principles of Green Chemistry

Although most of these pertain more strongly to chemical synthesis, these principles have been applied to the creation of the laboratories in this manual.

1. **Prevention**
   It is better to prevent waste than to treat or clean up waste after it has been created.

2. **Atom Economy**
   Synthetic methods should be designed to maximize the incorporation of all materials used in the process into the final product.

3. **Less Hazardous Chemical Substances**
   Wherever practicable, synthetic methods should be designed to use and generate substances that possess little or no toxicity to human health and the environment. Chemical products should be designed to affect their desired function while minimizing their toxicity.

4.

5. **Safer Solvents**
   The use of auxiliary substances (e.g., solvents, separation agents, etc.) should be made unnecessary wherever possible and innocuous when used.

6. **Energy Efficiency**
   Energy requirements of chemical processes should be recognized for their environmental and economic impacts and should be minimized. If possible, synthetic methods should be conducted at ambient temperature and pressure.

7. **Renewable Feedstocks**
   A raw material or feedstock should be renewable rather than depleting whenever technically and economically practicable.

8. **Reduce derivatives in synthesis**
   Unnecessary derivatization (use of blocking groups, protection/ deprotection, temporary

modification of physical/chemical processes) should be minimized or avoided if possible, because such steps require additional reagents and can generate waste.

9. Catalytic reagents (as selective as possible) are superior to stoichiometric reagents.
10. Design for Degradation
    Chemical products should be designed so that at the end of their function they break down into innocuous degradation products and do not persist in the environment.

11. **Monitor processes**
    Analytical methodologies need to be further developed to allow for real-time, in-process monitoring and control prior to the formation of hazardous substances.

12. **Prevent Accdents**
    Substances and the form of a substance used in a chemical process should be chosen to minimize the potential for chemical accidents, including releases, explosions, and fires.

*Anastas, P. T.; Warner, J. C. Green Chemistry: Theory and Practice, Oxford University Press: New York, 1998, p.30. By permission of Oxford University Press.

Responsible use of Fume Hoods

These labs will require minimal use of fume hoods. Fume hoods use energy and are based on the principle that "the solution to pollution is dilution." Analytical chemists know that this is not sustainable. Dilution can bring hazardous materials to a less toxic level but they are not gone!

In order to use fume hoods safely, please follow these guidelines from OSHA

1. Make sure that you understand how the hood works.
2. You should be trained to use it properly.
3. Know the hazards of the chemical you are working with;
refer to the chemical's Material Safety Data Sheet if you are unsure.
4. Ensure that the hood is on.
5. Make sure that the sash is open to the proper operating level, which is usually indicated by arrows or stops on the frame.
6. Make sure that the air gauge indicates that the air flow is within the required range.
7. Never allow your head to enter the plane of the hood opening. Keep the
sash below your face; for horizontal sliding sashes, keep the sash positioned in front of you and work around the side of the sash.
6. Use appropriate eye protection.
7. Be sure that nothing blocks the airflow through the baffles or through the baffle exhaust slots.
8. Elevate large equipment (e.g., a centrifuge) at least two inches off the base of the hood interior.

9. Keep all materials inside the hood at least six inches from the sash opening. When not working in the hood, close the sash.
10. Do not permanently store any chemicals inside the hood.
11. Promptly report any hood that is not functioning properly to your supervisor. The sash should be closed and the hood "tagged" and taken out of service until repairs can be completed.
12. When using extremely hazardous chemicals, understand your laboratory's action plan in case an emergency, such as a power failure, occurs.

If a lab uses safe materials, you do not need to use hoods. Use the benches. However, you must still follow safety rules and policies.

Washing glassware
Clean glassware sheets water. Be sure to wash glassware before use. First wash with soap and tap water. Follow by rinsing three times with DI/distilled water. Do not rinse with acetone! It's not necessary and could introduce contaminants.

Chemical Safety
Please become familiar with the Hazards diamond. A higher number (up to 4) indicates increased hazard. (http://www.ilpi.com/msds/ref/nfpa.html)

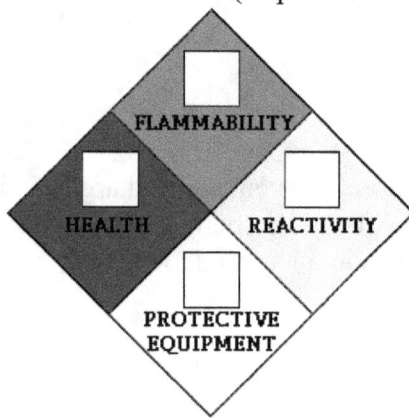

and
Safety Data Sheets (https://www.osha.gov/Publications/OSHA3514.html)

Categories of Chemical Reagents

- Technical (Commercial) grade is used when purity is not important. (Cleaning solutions for example.)
- Chemically pure "CP" chemicals are not as impure as technical but nature and amount of impurities aren't known.
- USP, United States Pharmacoepia, as found on vitamin supplements, means that the product doesn't contain chemicals hazardous to health but may be contaminated.
- Reagent Grade is the most common grade used in chemical laboratories. This grade is highly pure. Impurities and their concentrations are listed on the label.

- Primary standard grade is the highest purity, found with very pure and carefully analyzed reagents used to make standard or standardize solutions (solutions with known concentrations).
- Specific purpose chemicals are used for a specific method, spectroscopic or electrochemical for example.

Rules for Handling Reagents

- 1. Select the best grade of chemical needed and use the smallest bottle.
- 2. Replace the top of every container immediately after removal of the reagent.
- 3. Do not put caps and stoppers on the dirty bench top. Put caps down face up and hold the stoppers carefully between your fingers making sure to avoid contact.
- 4. Never put used reagents or ones that have been removed and put in a weigh boat or on weighing paper back in the bottle.
- 5. Never put a spoon or spatula in a bottle. Tap out what you need. If the reagent is caked, wash a spatula and dry it carefully before inserting into the bottle.
- 6. Clean up spills immediately.
- 7. You just want to analyze your sample. You do not want to analyze stuff in the water, leftovers form the last experiment or things left on the balance or lab bench. In Analytical Chemistry things need to be clean! Make sure to wipe down your lab bench with wet paper towels if it looks dirty. Check the balances. Wipe the top loaders with wet paper towels and dry them if they look dirty. If the analytical balances look dirty, notify the ta or me.
- 8. Wash glassware with the soap provided. Rinse with tap water and then rinse 3 times with our distilled/deionized water. Clean items should sheet water and not leave spots.
- 9. Wear gloves. Remove them before leaving the lab. Wash hands.
-

Central College Vermeer Science Center
Laboratory Rules and Policies

1. **NO HORSEPLAY IN THE LAB!**

2. Always look around to get a feeling for what others in the lab are doing, especially those in simultaneous classes when you are in VSC 263. Think about where you are standing, what you are touching, etc. Think about spills, explosions, hot glassware, sharps, etc. Be considerate of and watch out for others.

3. Know the location of the safety showers, eyewashes, fire alarms, fire blankets, fire extinguishers and phones. Do not obstruct aisles, hallways, or routes to safety equipment and building exits.

4. Keep long hair tied back. Do not wear sleeves that are loose at the cuff. Long pants, shorts that cover the knees, or a dress must be worn. Shoes must be worn, no sandals or other open shoes.

5. Chemistry students must ALWAYS wear safety glasses or goggles while in lab. You must provide your own. Biology and physics-check with instructor. Some labs do not require them.

6. Do not perform unauthorized experiments. Set up your experiment in the designated place for your lab and, if in VSC 263, do not encroach into the areas of other classes where you may put others at a safety risk.

7. Handle and store glassware with care. Do not use damaged glassware; exchange it at the stockroom. Dispose of broken glassware and disposable pipettes in the appropriate "Broken Glass" cardboard boxes.

8. All sharps (needles, scalpels, etc.), or anything that has touched blood or bodily fluids **and** has a potential to be a sharp if broken, e.g. a microscope slide used for blood typing, are to be disposed of in the red plastic sharps containers located on the lab benches. The cardboard "Broken Glass" boxes are NOT for glass with blood on it. Do not place tissues or other trash in the red plastic sharps containers or in the cardboard broken glass boxes. Unless a tissue is very wet with blood, it can be placed in the regular trash. Also, always wear gloves when working with blood, saliva, urine, or any other bodily fluids. Wash hands and change your gloves between "patients" and use a disinfectant on your work area.

9. In biological labs where infectious substances are used, such as microbiology, practice good hygiene and organization of your work area. Follow your instructor's guidelines for cleaning your work area and washing your hands to prevent cross-contamination.

10. Know the physical hazards and health hazards of the chemicals or substances you are using. Ask your instructor or the stockroom manager or search the literature. Material Safety Data Sheets (MSDS) as well as the Merck Index are available at the stockroom. The science reading room has resources as well. If you feel you must work with a substance before you are sure of its hazards, work in the fume hood while wearing gloves and eye protection.

11. Never work alone in the lab. An unfortunate accident could quickly incapacitate you. For example, you may not be able to get yourself to the eyewash in case of a chemical spray or splash to the eyes. Another example: being overcome by fumes with no one available to call for help for you.

12. Label **every** container that you put something into, using lab tape or a grease pencil, with the following; the **date** including the year, the **contents** with approximate **amounts**, and your **initials**. *<u>If a container is not properly labeled, its contents will be disposed of at the end of the day</u>*. Never use substances from unlabeled containers. Give all unlabeled substances that you may find to the stockroom for analysis and disposal.

13. Always dispose of chemicals by using the appropriate containers. If in doubt of which container to use, put the substance in its own container, seal it properly, and label it with as much information as you know and give it to the stockroom or your instructor.

14. Never taste chemicals. Smell a substance only if you know it is safe (ask, or search the literature), and be sure to waft. Do not pipette or siphon by mouth; use bulbs or syringes.

15. Eating, drinking, chewing gum, applying cosmetics or touching contact lenses should never be done in a lab room. Always wash your hands thoroughly with soap and water when finished with lab and before going to the restroom or performing the activities mentioned above.

16. When using the fume hood, be sure you are protecting yourself by keeping the sash at the smallest opening in which you can work comfortably. Keeping the sash all the way open

defeats the purposes of the hood, which are to provide a face/body shield and to provide high velocity exhaust. Always perform work 6 inches or more back from the front of the hood.

17. Immediately notify the instructor, TA, stockroom assistant or stockroom manager of the following; any spill larger than about a beaker (100 mL) of any substance, any size spill of a toxic substance, slick walking surfaces, broken glass, broken mercury thermometers, fumes or bad odors, health symptoms (headache, etc.), any inhalation or exposure of eyes or skin to a substance, any fire, or any violent chemical reaction.

18. **Clean up your work area when you are finished.** Clear off and wipe down the counter top. Put everything away and make sure you have labeled any container in which you are keeping something. If you have an experiment in-progress, make sure it will be safe (use a hood if necessary) and fill out a card to be placed with your experiment with **your name**, **your course**, the **date and time**, the **experiment** name, and any **special precautions**.

19. **DO NOT remove any laboratory items; instruments, consumables, chemicals, from VSC. And DO NOT** touch, move, or disrupt lab items on tables when in lab, especially Intro lab 263 weather you are registered in that lab or not unless permission is given by your instructor.

20. **Stay out of the Stockroom**! If you are in a scheduled lab, any business you have with the stockroom can be conducted at the rollup window, with the on-duty stockroom assistant. If you are doing lab work at a time when no labs are scheduled, you may not be afforded the same access to supplies. Also, any labs done after **4 PM** are to be arranged with your instructor. For independent labs, if you plan ahead and give a list of needed supplies to the stockroom manager one week ahead of time then everything can be found and a special tray or other space can be assigned for your supplies. Do not wait until the last minute and expect to have everything that you need. If you are a lab TA, a lab set-up person, or a Stockroom assistant, you should **not** be in the Stockroom without permission **unless** you are on duty.

USING THE STOCKROOM

Through the Vermeer Science Center Stockroom you can obtain all the necessary supplies, equipment, and other resources to do your lab work. The Stockroom is the primary department for these things, but none of these resources can be used if you have not signed the VSC lab rules and policies agreement for the current term, or if you have outstanding lab charges, or if you are a habitual violator of VSC lab rules and policies.

The basic guideline for using stockroom resources for supplies, equipment, glassware, chemicals, and some special items is...
> Use the sign out sheet on the clipboard that is on the counter top at the rollup window. Fill out your ID#, Name, Item(s) requested (one item per line) and the date. The stockroom assistant will get your item(s) for you. When you turn the item in (**all glassware is to be clean**), make sure the stockroom assistant checks the item(s) as returned or crosses it off the sheet.

For requests to order special chemicals or equipment for lab work you are doing, see the Stockroom manager.

The Laboratory Notebook
1. Put a **title** on your lab. Sign and date each page. Write in the notebook using a pen. The title should include the method used, the analyte, and the sample.
2. If you have a **sample number** it should go right below the title. State how the sample was obtained and any observations.
3. Write a short **introduction** in which you mention the basis for the analysis and any important chemical equations.
4. Include **Hazards**: review each chemical and its safety ratings.
5. **Methods and procedure**: Outline the steps you take in the lab and the observations you make. Unless this is a formal typed report, you should be writing this in your lab book as you do the lab, not after. Do not simply list the steps from the book but write what you did. Record procedures for making all of your solutions. Include the number of your balance and the make and model of any instruments that you use. Record all data. Make sure you record all units. Include observations.
6. **Data and data analysis**: collect data during the lab. It's a good idea to do some data analysis to make sure your data is good before you leave the lab.
7. Show a **sample calculation** for each calculation that you need to do. Include all **tables and graphs**. Each table should have a title at the top of the table. Graphs should be labeled as Figures and also have a caption. Be sure to keep track of significant figures.
8. Report the **results** and accuracy (if known) and precision of your analysis. Discuss concepts learned, modifications that you'd make in the lab if you were to repeat it. If there are questions in the laboratory procedure, answer them here.
9. **Conclusions.** Restate your results, including your accuracy and precision and sample identification/number. Use your results to come to your conclusion and refer to them.
10. Don't obliterate mistakes. Cross them out with a single line.

Use of the Balance
1. If you want more significant figures you will want to use an analytical balance. If you want fewer significant figures, use a top loading balance in the lab.
2. Make sure the balance is clean before and after you weigh.
3. Do not handle objects to be weighed with your fingers. Handle with tongs or a clean piece of paper or finger cots.
4. Use a weighing bottle to weigh solids. Some solids will need to be dried before weighing. But don't weigh a hot object. Weighing by difference is often the easiest way to get a measurement.
5. If you use weighing paper, do not use an analytical balance.
6. Close the balance doors and make sure the balance is level before weighing. Don't lean on the balance table.
7. A balance is not a scale. A balance compares masses, a scale measures mass based on gravitational force.

Experiment 1: Lab check in, balances and desiccators

Background:
If you want four or more significant figures in your measurement of mass you will want to use an analytical balance. If you want fewer significant figures, use a top loading balance in the lab.

A few rules for using a balance:
1. A balance compares masses. A balance is not affected by gravity. A balance is different from a scale. Please don't call it a scale.
2. Make sure the balance is clean before or after you weigh.
3. Do not handle objects to be weighed with your fingers. Handle with tongs or a clean piece of paper.
4. Use a weighing bottle or a glass weighing funnel to weigh solids. (Weighing paper might hang over the edges of the balance pan and make weighing difficult.) Some solids will need to be dried before weighing. But don't weigh a hot object. Convection currents caused by the heat will affect the weight. Weighing by difference is often the easiest way to get a measurement.
5. Close the balance doors and make sure the balance is level before weighing.

A desiccator is used to keep objects dry. The lid of a desiccator should slide easily and the desiccant should be blue. Always hold the lid when moving the desiccator. Place hot objects in the desiccator using tongs and vent the lid three times before closing.

Procedure:
In your lab drawer, locate three glass filtering crucibles or weighing bottles. Weigh each on an analytical balance. Handle with tongs. Record the weight in your lab notebook. Place one crucible in the drying oven at 110^0C for 30 minutes. Don't handle it. Let the other one sit out in the air. Let the third sit in air and handle it with your fingers. Check in the remainder of your drawer and locate your desiccator. Retrieve your crucibles/bottles. Allow the one heated in the oven to cool in your desiccator for 20 minutes while you weigh the other two. Did the weight change? What % of the weight changed?
Review: (Weight after-weight before)/ weight before x 100= % change
Summarize your results and compare with those of two other students.
What can you say about the difference in your observations?

Experiment 2a: Observations of a Candle Burning

Background:

Born poor, Michael Faraday wanted to make sure that all children found science wondrous and accessible to them. He gave lectures to young people that became known as the "Christmas Lectures" covering the fields of chemistry, electricity and magnetism, gravitation, as well as other topics in physics.

THE CHEMICAL HISTORY OF A CANDLE, by Michael Faraday published around 1860, was based on a lecture given by the famous scientist. The idea behind this experiment is to enhance important qualities of scientists:
1. Curiosity about the natural world
2. The ability to observe open-mindedly and infer
3. The ability to communicate with others.

Review:

Keep long hair tied back. Do not wear sleeves that are loose at the cuff. Long pants, shorts that cover the knees, or a dress must be worn. Shoes must be worn, no sandals or other open shoes. Chemistry students must ALWAYS wear safety glasses or goggles while in lab

Procedure:
1. Weigh your candle on an analytical balance (in the weighing room) and record mass.
2. Put candle in the hood.
3. Describe the candle while it is not burning.
4. Light the candle carefully.
5. Let the candle burn for about 10 - 15 minutes making all of the observations you possibly can. Write your observations down.
6. Place a beaker over the candle. Continue to observe the candle. Write your observations down.
7. Reweigh the candle. (Make sure it has cooled before you do this.)

Further Investigation (Work with a group for this part):
1. Cut a piece of wick provided (approx. 2 inches). Place one end of the wick in a dish of water. Hold the wick in the water with tongs. Light other end of the wick. Record your observations.
2. Place the same wick in a dish of lamp oil or kerosene or preferably a small oil burner. Light it. Record your observations. Have a clean beaker ready to smother the flame when you are finished.
3. What conclusions can you make? What type of substance would be best to make a lamp or candle burn?
4. Based on your observations, propose a chemical equation to explain the chemistry of a burning candle.
5. Do some research online. What is the chemical equation for a candle burning?
6. Make a conclusion about the candle burning and your proposed formula.
7. Using the equation found online and the loss of weight of the candle, calculate how much carbon dioxide was created during the burning of your candle.

Experiment 2b: Sampling
Lab Created by Zachary M. Miller with Cathy Haustein

Below is a grid in which dots are distributed unevenly. Count all the dots.

	1	2	3	4	5	6	7
A	::	:	.	.	:::	.	
B	:	:	.	:	:	.	.
C	.	.	::	::	::	:	:
D	:.	:	:	:	.	.	
E		.	.	:.	:.	:	.
F	:	.	:	:	:	:	.
G	.	:		::	.	:.	:

1	2	3	4	5
6	7	A	B	C
D	E	F	G	

1. Cut apart the above numbers and letters Discard the blank square. Make a pile of numbers and a pile of letters, face down. Shuffle each pile separately to randomize the two piles.
2. Pick one letter and one number. Locate that area on the grid and write down the number of dots. Multiply this number by 49 to estimate the total number of dots in the grid based on one sample. The number and letter do not go back in the piles.
3. Repeat step 2 six more times for a total of seven trials.
4. Add the number of dots in each trial together, then multiply the total number of dots by 7 to estimate the number of dots based on seven samples.
5. Pool class data and average for the first trial in step 2.
6. Pool class data and average for the seven trials from step 4.
7. Calculate the absolute and relative percentage error for each set of trials. (measurement-true value)/true value x 100=% relative error
8. What can you conclude about sampling and a non-homogeneous sample?

Experiment 3: Using a Spreadsheet, Buret, and Pipet

Background: We will use a spreadsheet to calculate the average weight of ten milliliters of water. Then we will convert this to mL using density. Standard deviation will be calculated as a way to determine precision. Each person in the class should do two trials with a pipet and two with a buret and the class will pool data using a spreadsheet. Students will compare the accuracy and precision of the methods.

Definitions;
Aliquot is a carefully measured portion
Accuracy is closeness to known value
Precision is reproducibility.

Review:
1. Buret
Make sure tip is filled with water. Make sure it is level. You will want to read to the 100ths place. Read meniscus.
2. Pipet
Read meniscus. Keep finger dry.
TD to deliver. Don't blow out last drop. Hold at an angle when delivering. Touch side of pipet to container for three seconds after its empty to make sure it has drained.
Most of our pipets are TD.
Blow out last drop with bulb. If you get liquid into the bulb, wash it out with isopropanol.
3. Weighing
Don't add weight from fingers. Use finger cots or paper to manipulate. Keep cork on to prevent evaporation! Make sure balance doors are closed. Don't lean on counter or weigh hot objects.
4. Accuracy is nearness to the true value. For this lab, the true value is 10.00 mL
5. Precision is nearness of the results to each other. We will determine this using the standard deviation.

Procedure:
1. *Experiment.*
    1. Obtain a small dry flask and cork.
    2. Weigh it on an analytical balance.
    3. Deliver a 10 mL aliquot with a pipet. Quickly cork and reweigh.
    4. Use the thermometer provided record the water temperature.
    5. Build a spreadsheet to calculate the weight of the water for several trials.
    See step 2 for information.
f. Repeat with 10.0 mL DI water added with a buret.
g. Use density of water and temperature (be sure to measure it) to calculate the milliliters delivered by each method. Make sure to look up the density of water at your temperature and to measure the water temperature used for the pipet and buret separately.

Are they equally accurate? What would you expect from knowing the tolerance of each measuring device?

h. STDEV is standard deviation, a measure of precision. A small value means that the measurement is more precise. *Which method of delivering water is more precise?*
i. Repeat using a buret.

Condensed version of a how to make spreadsheet.

. Open the program Excel
a. Create 3 columns
   1. Click on A1 and type in "with water"
   Click on B1 and type "empty"
   Click on C1 and type "grams water"
b. Add data of the weight of the beaker, cork and water to A2, A3, A4 etc.
c. Add data of the weight of the beaker and cork to B2, B3, B4 etc.
d. Add a formula in C2.
   Formulas are indicated by =() Type in =(A2-B2). Then copy from C2 to other cells in C.
e. Add data from the class.
   In a cell not used up by data in C type =AVERAGE(Cx:Cy)
   For values of x and y either highlight the cells or type in the numbers of the row.
   Write "average" in row D.
e. In another cell in C type =STDEV(C#:C#). Put in the numbers or highlight columns.
f. Alternatively, calculate the mL of the water directly and enter into a spreadsheet.

Results. Record the absolute error (10.00 mL -experimental value) and the standard deviation of each method of measurement. Combine the class data.
Calculate the confidence interval at the 95% level.

Conclusion. Discuss if your results are reasonable and to be expected. Restate your results, including your accuracy and precision and sample identification.

Experiment 4: Gravimetric Determination of Phosphorus in Plant Food Using Household Chemicals

Based on a lab by Sally Solomon, Alan Lee, and Donald Bates, Drexel University, Philadelphia, PA 19104

Also : http://www.esu.edu/~scady/Experiments/Fertilizer.pdf

Background:

This is a chemical method for the determination of phosphorus in plant foods is using a gravimetric method. It requires no special waste-disposal methods. A precipitate of magnesium ammonium phosphate hexahydrate is formed. The molar mass of this is 245.4070 g/mole. Atomic mass of phosphorus is 30.973762 g/mole. Molar mass of P_2O_5 is 141.944522 g/mole

Plants require essential macronutrients that may be supplied by fertilizers. The three important nutrients sometimes depleted in soil are called primary nutrients: nitrogen, phosphorus, and potassium. Nitrogen is used for leaf growth while phosphorus and potassium help fruit and root growth. Plant food labels display numbers in the order N-P-K that indicate their percentages and the form of the nutrient used to calculate the percentages.

        5% N   4 % P_2O_5 15% K_2O

Is indicated by 5-4-15

The gravimetric determination is based on the precipitation of magnesium ammonium phosphate hexahydrate in a basic solution.

$$5H_2O(l) + HPO_4^{2-}(aq) + NH_4^+(aq) + Mg^{2+}(aq) + OH^-(aq) \leftrightarrow MgNH_4PO_4 \cdot 6H_2O(s)$$

The weak base ammonia will raise the pH without forming insoluble $Mg(OH)_2$.

Chemicals:

    All available as household products.
1. Rubbing alcohol (isopropyl alcohol, 70% or greater concentration)
2. Magnesium sulfate (Epsom salts)
3. Ammonia (ammonium hydroxide)

- **Review: Before lab, write a summary of what the lab will be about. List all important equations.**
- **List all chemicals that you will use and the hazards classification of each one.**
- **A sample should be homogeneous and represent the whole.**

Procedure:

For a solid sample,* grind a little over ten grams in a mortar and pestle. Weigh the sample of plant food to the nearest 0.01 g. Dissolve the plant food in 125-140 mL of DI water. Stir for five minutes with a glass stir rod. Filter into a 600-800 mL beaker to remove insoluble material. Fold filter paper in fourths, rip off one corner, and place firmly in your funnel as shown here:
http://community.asdlib.org/imageandvideoexchangeforum/2013/07/24/gravity-filtration/
Add 150 mL of a solution of 10% w/v $MgSO_4 \cdot 7H_2O$ to the sample solution (filtrate) with stirring. The $MgSO_4 \cdot 7H_2O$ solution used for precipitation should be prepared by dissolving 15 g of Epsom salts/150 mL water. Each analysis requires about 150 mL of Mg^{2+} solution.

To the plant food filtrate (or to the liquid plant food), add approximately 180 mL of NH_3(aq) (commercial product). If commercial product not available, 60 ml of 20% ammonium hydroxide may be used. Add the solution **gradually** while swirling or stirring. A white precipitate then forms.

Let the suspension stand at room temperature for 5 minutes and in an ice bath for ten minutes. Filter with Whatman 40 or 49 paper in a 100-mm funnel. To hasten filtering, decant off as much of the liquid as possible through the paper before adding the solid. A Buchner funnel may be used but it wastes water. Notify the instructor if the precipitate doesn't filter within a half hour. A hard to filter precipitate forms if the ammonia is added too quickly.

Add 20-mL of rubbing alcohol to the precipitation flask to remove any remaining material. Pour over the solid in the funnel to aid drying. Repeat.

Remove paper and precipitate from funnel. Spread it out and place it on a flat surface such as a watch glass or paper plate (for faster drying). Break up lumps with a spatula. The precipitate should be left to dry overnight or longer.

Scrape the solid off the filter paper into a weighing dish, then weigh. It's not possible to find the weight of the precipitate by subtracting the weight of the dry filter paper since the later absorbs water. Clean lab area with a wet paper towel.

Calculations and Write-ups:

From the weight of the $MgNH_4PO_4 \cdot 6H_2O$ calculate the percentage phosphorous in the plant food sample. To check your results for accuracy compared to the label, convert the %P to % P_2O_5.
Be sure to verify this using the gravimetric factor.

For your write up, pool class data to come up with a class average for %P and % P_2O_5, then evaluate your own data. Keep in mind that products are often overformulated which means that there is often extra product. In general, plant food contains about 1% higher P_2O_5. Collect data from the entire class and calculate the mean, median, standard deviation and relative standard deviation (as a percent). You should be within 1.3% of the value of the class. If not, it is likely that you had a determinate error. Discuss your error in your conclusions.

* If a liquid plant food is used pipet a 10.0 mL -15. 0 mL sample.

Experiment 5: Titration of KHP
Adapted from: Analytical Chemistry Sixth Edition by Gary D. Christian

Background:
This lab will introduce you to a simple neutralization titration using a strong base as the titrant and an unknown containing KHP (this stands for potassium hydrogen phthalate). The titrant, sodium hydroxide, can't be made directly and will be standardized. The w/w% KHP and precision will be reported.

Hazards:
Look up and record the hazards for KHP, NaOH, and phenolphthalein. Note: other names for KHP are potassium biphthalate, potassium hydrogen phthalate potassium, acid phthalate and phthalic acid, potassium salt.

Review:
- Blue Health
- Red Flammable
- Yellow Instability
- White Specific Hazard
- OXY (oxidizer like bleach,) ACID, ALK alkaline or basic), CORR (corrosive), (also use no water and radiation hazard)
- Higher number means more precaution is needed.
- Make sure to wipe down your lab bench with wet paper towels if it looks dirty. Check the balances. Wipe the top loaders with wet paper towels and dry them if they look dirty. If the analytical balances look dirty, notify the ta or professor.
- Wash glassware with the soap provided. Rinse with tap water and then rinse 3 times with our distilled/deionized water. Clean items should sheet water and not leave spots.
- Wear gloves. Remove them before leaving the lab.
- Clean up all spills. Spills left unattended will result in lost lab points.

Equation: http://www.slcc-science.org/chem/labs/chem1115/NaOH/NaOH.htm

Simplified version: $HA^- + OH^- \rightarrow A^{2-} + H_2O$

Carbon dioxide in the water will produce a weak acid so carbon dioxide must be removed. Complete this reaction and add it to your notebook. The weak acid H_2CO_3 acts as a buffer and obscures the endpoint

$CO_2 + H_2O$

Procedure:
1. *Provided.*
    KHP primary standard grade
    1. 0.2% phenolphthalein solution in 90% ethanol
    50% NaOH solution. Hazardous. Use gloves and eye protection. Rinse any excess in the sink with plenty of water.
2. *To prepare the titrant,* 0.1 M NaOH solution. To make sure it is carbonate free, begin with 50% NaOH solution (this will be provided to you).

Distilled water free from carbon dioxide will be needed. To prepare this, boil the water for a few minutes. Fill a 1000 mL beaker with 1000 mL of DI water, insert a glass stir rod to minimize the chance of it boiling over, heat to boiling on a hot plate, and boil for 5 minutes. Cool in an ice bath.

Fill a 1 L plastic bottle with the CO_2-free water, add 5-6 mL clear 50% NaOH solution, cap, and swirl to mix. Avoid exposing the solutions to the atmosphere as much as possible. Cap, and shake thoroughly but carefully. Label as 0.1 M NaOH.

Review:
- Reagent Grade. A highly pure chemical. Most common grade. Impurities and their concentrations are listed on the label.
- Primary standard grade. Highest purity. Very pure and carefully analyzed. Used to make standard solutions (solutions with known concentrations).

Do this while you are waiting for your water to boil.
Dry about 5-6 g primary standard KHP in a weighing bottle at 110-120°C for 1 hour. Be sure to dry with caps off!
Cool in a desiccator for at least 30-40 minutes before weighing. Dry in a labeled beaker that is covered with a watch glass. At the same time, select a solid unknown. Be sure to record the sample number and observations about the sample in your notebook. Dry for 1 hour at 110-120°C along with your pure KHP. Store in the desiccator. Be sure to handle with tongs and vent the lid of the desiccator three times.

Procedure for titration:
Heat 500 mL of water to boiling and allow to cool. This will be used to dilute the titrations.

Standardization of the 0.1 M NaOH solution this way:
1. Using an analytical balance, weigh accurately (using the analytical balance) three portions of the dried KHP of about 0.5 g each, and transfer to clean 250-mL wide-mouth Erlenmeyer flasks. The direct method of weighing, using a tared weighing funnel or boat should be used with this material.
2. Dissolve each sample in about 50 mL CO_2-free distilled water.
3. Rinse your buret with three small portions of the 0.1 M NaOH solution, fill, and adjust to near zero. Record the initial volume reading to the nearest 0.02 mL.
4. Add 2-3 drops phenolphthalein indicator to each KHP sample and titrate with the 0.1 M NaOH to a faint pink end point. The color should persist at least 30 seconds. Split drops at the end of the titration. Estimate the buret reading to the nearest 0.02 mL.
5. Calculate the molarity of the NaOH to four significant figures from the weight of KHP used. Use the average of the results. Remember that KHP is an abbreviation for an organic acid. It has a molar mass of 204.22 g/mol.

Determination of KHP in an impure sample

Since the unknown contains KHP you will use phenolphthalein indicator (color change same as above).

Weigh three samples of .9-.8 g to the nearest 0.1 mg (0.0001 g) into Erlenmeyer flasks. Dissolve in 50 mL CO_2-free water. Add two drops of indicator and titrate with 0.1 M NaOH until the color change persists 30 seconds.

Calculation:

Calculate and report the w/w percent KHP in your unknown based on the average of the three trials, or report the median.

Rinse and refill buret with DI water.

Review:

Show a sample calculation for each calculation that you need to do. Be sure to keep track of significant figures.

Report the results and accuracy (if known) and precision of your analysis.

Restate your results, including your accuracy and precision and sample identification in the conclusion.

Experiment 6: Titration of vinegar, wine, and soft drinks

Background:

Titrations are inexpensive and relatively quick chemical methods. In this lab we will determine the amount of acid in two consumer products. You might have to adjust the sample sizes here to make sure that you do not need to refill your buret and that you deliver over 10.0 mL of titrant. Other than that, this lab is similar to the previous lab. Use your standardized NaOH as the titrant. For calculation purposes we will assume that the acid in vinegar is acetic acid and the acid in wine is tartaric acid.

Prelab:

Do background research on acids in soda, vinegar and wine. Have this information in your "Introduction" section of your lab.

Equations:

In your prelab, indicate your neutralization reactions. Keep in mind that tartaric acid has two protons and will need two moles of NaOH to neutralize it. Citric acid has three titratable protons.

Review: Purely chemical methods were developed in the nineteenth century and therefore are called classical methods. Although they are old methods, they are highly accurate and precise.

For a TD pipet, hold tip to side of volumetric flask to drain. Keep there for three seconds after all liquid has been delivered. If you get liquid in the bulb, wash it out with isopropanol.

Procedure:
1. *Titration of Vinegar.*
    1. Take a 5.0 mL aliquot of vinegar.
    2. Add to an Erlenmeyer flask and dilute to 50 mL with DI water. Add three drops of phenolphthalein.
    3. Titrate quickly with standardized NaOH to the phenolphthalein endpoint.
    4. Repeat. (Adjust the amount of vinegar if needed.)
    5. Repeat two more times for a total of three careful titrations.
2. *Titration of Wine.*

a. Use a 10.0 mL aliquot of wine for the first trial. Put this directly into your Erlenmeyer. Dilute to 50 mL with DI water.

b. Titrate to the phenolphthalein end point as before. Make sure to get three good trials.

It might not be possible to get a good phenolphthalein endpoint and you may need to adjust your aliquot size. Repeat two more times. Based on what you have read about acids in wine, are your results reasonable? Comment on the selectivity of this titration.

3. *Titration of Soft Drinks*
    a. Use a 15.0 mL aliquot of a clear soda. Titrate as before, assuming that the acid is citric acid.

Calculations:
1. Determine the w/v% acid in vinegar. Is this what should be expected?

2. Determine the w/v% of tartaric acid in wine and the range of values at 90% confidence. Is this what should be expected?
3. Determine the w/v% citric acid in the soda.
4. Remember to refill your buret with DI water at the end of this lab.

Experiment 7: Acid-base titrations with plotting and unknown acid

Goals:
1. To help you become familiar with weak acid/strong base titration curves by creating a curve for KHP and for an unknown acid.
2. To give you experience applying a method to a real life situation.
3. To incorporate citations into a lab report.

Background:
    Neutralization titrations are a classic method of analysis. They are inexpensive and relatively quick. Creating a plot mL titrant vs pH using a pH meter and a glass electrode will generate a titration curve. This can reveal a lot about the substance titrated. If a weak acid or base is titrated and a titration curve is made, the molecular weight and Ka or Kb can be calculated.

Review:
    Wash glassware with the soap provided. Rinse with tap water and then rinse 3 times with our distilled/deionized water. Clean items should sheet water and not leave spots. Wear gloves. Remove them before leaving the lab.

Sketch the titration curve you expect to see below, indicating the midpoint and the equivalence point:

Procedure:
1. *Titrations.*
    1. Titrate a monoprotic acid and plot the results.
        1. Dissolve 2 g of dried KHP (to 4 significant figures) in a 100 mL volumetric flask; dilute with DI water.
        2. Titrate a 15 mL aliquot, using phenolphthalein as an indicator.
        3. Add 25 mL DI water before titrating to add volume to your sample.

b.   Repeat the titration following the pH changes, use a pH meter and glass combination electrode.

c.   Calibrate the pH meter with the pH 4 and pH 10 buffers provided. Record the initial pH. As the titration progresses, record the mL of NaOH added, the pH, the (first derivative) change in pH/ change inmL NaOH in your lab notebook. Add 2 mL initially, then titrate in 1 mL increments until you approach the midpoint
Use your previous titrations to estimate the midpoint.
Add NaOH in 0. 5 mL increments near the midpoint.

After the theoretical midpoint, increase the amount added until you are within 1-2 mL of the endpoint.
Once there, add titrant one drop at a time.
Make sure that the solution is stirred completely and take a reading after each drop.
After the endpoint, add an additional 1 mL and then 5 mL of NaOH and record the pH.
Note when your indicator changes color.
Make a plot of pH vs mL NaOH and also of the "first derivative" vs mL NaOH.
Use the known Ka value and molecular weights, evaluate your accuracy.
This should be attached to your final report as an appendix.

The Unknown.
   .   Repeat steps above using an unknown acid.
           1. <u>However</u>, use 1-1.5 g of acid in 100 mL.
Some of the acids are only slightly soluble in water. Watch your acid carefully and if it looks as if it is not dissolving, add some ethanol to the volumetric flask.
           2. If this is a diprotic acid with careful titration you might be able to record two pH jumps for part 2 if the K_a values are far enough apart.
a.   Make a plot as above
           1. Use the known K_a values and molecular weights, identify your acid.
b.   Repeat titration and plot, so that you can evaluate your deviation.

Application.
   .   Identify a consumer or other product that contains your substance.
a.   Titrate that product and determine the amount of your acid in that product.
b.   Do a literature search about your acid using the ACS web link on the library page. Also search other databases.

Experiment 8: Acid-base titration plots using known acids.

Background:
Neutralization titrations are a classic method of analysis. They are inexpensive and relatively quick. Creating a plot mL titrant vs pH using a pH meter and a glass electrode will generate a titration curve. This can reveal a lot about the substance titrated. If a weak acid or base is titrated and a titration curve is made, the molecular weight and Ka or Kb can be calculated. This lab will help you become familiar with weak acid/strong base titration curves by creating a curve for KHP and for maleic acid. (For a less toxic diprotic acid, tartaric acid may be used but the two endpoints will be less visible.)

Don't forget: find and list hazards for maleic acid.

Titrations

1. In the first part of the lab you will titrate a monoprotic acid and plot the results. For the monoprotic acid, use some previously dried KHP. Dissolve about 1.5-2 grams (weighed on an analytical balance) in 100 mL of water using a volumetric flask. (If you have a weighing funnel, use this instead of a plastic weigh boat.) Titrate a 20 mL aliquot using phenolphthalein as an indicator. Add 25 mL di water before titrating to add volume to your sample.

2. Now repeat the titration following the pH changes with a pH meter and glass combination electrode. Titrate using the pH meter to record how the pH changes as you add NaOH. Make sure to calibrate the pH meter with pH 4 and 10 buffers provided. As you titrate, follow the Δ pH/ Δ mL NaOH and record this in your lab notebook. Add 2 mL initially, then 1 mL increments until you approach the midpoint. (Use your previous titrations to estimate the midpoint.) Then add in 0.5 mL increments near the midpoint. Once past the theoretical midpoint, increase the amount added until you are within 1-2 mL of the endpoint. Then add titrant one drop at a time. Make sure that the solution is stirred completely and take a reading after each drop. Make a note of where the phenolphthalein changes color. (This is the indicator endpoint.) Once you are past the meter endpoint (where there has been a large distinct pH change), add an additional small portion of NaOH, followed by 5 mL of NaOH. Then add another 5 mL.

3. Make a plot of pH vs mL NaOH. Using your data, calculate the Ka and molar mass (molecular weight) of the acid. Use the known Ka value and molecular weight, evaluate your accuracy. This should be attached to your final report as an appendix. (If instructed to, make a first derivative plot as well.) Repeat.

Next. Repeat steps 1 and 2 using maleic acid. However, use about 1.2 g of acid (weighed carefully) in 100 mL. This is a diprotic acid and with careful titration you might be able to record two pH jumps for part 2. Make a plot as in Step 3. Use the known Ka values and molecular weights, identify your error. Repeat so that you can evaluate your deviation.

Experiment 9: Natural Indicators

Background:
Neutralization titrations are a classic method of analysis. They are inexpensive and relatively quick. A titration can be used to determine the molar mass of an acid or base if the appropriate indicator is used.

Phenolphthalein was once used as a laxative but now is a suspected carcinogen. It is no longer used in laxatives but we still use it in lab. Are there any natural products out there that can be used instead? One substance used as a natural pH indicator is red cabbage.

Indicators are weak acids or bases which change color at a specific pH. This should be at or near the equivalence point. Indicators show a change in pH comparable to $pK_a \pm 1$.

Reminder: Be sure to record good observations of color changes. Carefully record your procedures for creating the indicators.

Procedure:
1. *Preliminary work.*
    1. Prepare an indicator solution by making an extract of red cabbage. You may boil a portion of the cabbage or soak it in alcohol. The second option will be faster and less smelly. (Record your own procedure and observations but use a large leaf with 50 mL rubbing alcohol.)
2. *Titrations.*
    1. In this part of the lab you will titrate a monoprotic acid and a diprotic acid and determine the molar mass. From this, you will evaluate three indicators. For the monoprotic acid, use some previously dried KHP if available.
        1. Dissolve about 2 grams (weighed on an analytical balance, using your glass weighing funnel) KHP in 100 mL of DI water using a volumetric flask.
        2. Titrate a 10 mL aliquot using phenolphthalein as an indicator. Add 25 mL DI water before titrating to add volume to your sample. Follow another titration using a pH meter. Note the pH where the indicator changes color. You do not have to generate a complete titration curve. Repeat.
        3. Repeat with the diprotic acid tartaric acid.
        4. Use cabbage indicator in place of phenolphthalein. Again, titrate once and then follow two titrations with the pH meter to note where the indicator changes color. (You do not have to generate a titration curve.)
        5. Compare the results of the cabbage indicator with that of phenolphthalein. To do this, use your data to calculate the molar mass of the KHP and compare it with the actual molar mass. Repeat for tartaric acid, again noting the pH where the color change occurs.

6. Next, identify another substance that might be a suitable indicator. Make that indicator and evaluate as previously done. Use the molecular weights, evaluate your accuracy. Be sure to note the pH at which the color change occurs and the details of how you made your indicator.
7. Write a strong conclusion, explaining why the indicators did or did not give accurate results. Evaluate precision. Include a discussion of the molecules that change color with pH for each indicator.

Experiment 10: Buffering Action of Hair and Skin Products

Adapted from: Analytical Chemistry Sixth Edition by Gary D. Christian

Background:
Hair and skin contain proteins and proteins are sensitive to pH. In this lab you'll use a calibrated glass pH reference to measure the pH of hair and skin products, both concentrated and diluted 1:10. We will compare this with the pH changes seen in unbuffered solutions. From this, decide if pH is controlled primarily by a buffered solution or a strong electrolyte. In your conclusions, evaluate the products as to buffering ability and discuss how pH affects hair and skin.

Equations:
Electrode Response
$$E = k - 2.303RTFpH \qquad E = k - 0.0592\,pH$$

Review:
pH is the $-\log[H^+]$

A buffer is resistant to changes in pH due to dilution. Buffers are made from a weak acid its conjugate base or a weak base and its conjugate acid in the same solution.

$$\beta = 2.302\, C_{HA}C_A/(C_{HA}+C_A) \quad \text{Units are mole/L per pH unit}$$

Reagents:
1. *Provided by professor.*
   Standard buffers and consumer products plus 0.1 M HCl. (You are welcome to bring a sample of your own product to test.)

Procedure:
The experiment will be performed by first calibrating the electrodes.

Prepare a concentrated sample (about 5 mL) in a 10 mL beaker. This will be just enough to cover the bulb of the glass electrode.

Prepare a 1:10 dilution of each sample in a 100 mL beaker by diluting 5.0 mL with 45.0 mL DI water.

Measure the pH of the concentrated sample (product) and then the diluted sample. Immerse the electrode in the test solution and swish or agitate a few seconds. Then allow the pH reading to equilibrate and record to the nearest 0.01 pH. Rinse the electrode well between measurements and blot off the water.
Repeat each trial. Repeat with four more products.

For comparison with an un-buffered solution, measure the pH of 5.0 mL of 0.10 M HCl. Dilute it to 50 mL and re-measure. Repeat with a solution of vinegar. Prepare a solution of sodium acetate from .5 g in 50 mL of water. Repeat the pH and dilution measurements as before. Mix the vinegar and sodium acetate. One again, measure the pH of the solution, then of the diluted solution.

Report: Prepare a table with these categories:
pH undiluted, pH diluted, Δ pH, buffer (yes or no)

Consult the reference J. J. Griffin, R. F. Corcoran, and K. K. Akana, *J. Chem. Ed.*, **54** (1977) 553* and include in your report a discussion of the relevance of pH in hair cleansing or conditioning and in hair damage. Propose the buffer acid base conjugate pair.

Also discuss which products have the most and least buffer capacity based on the magnitude of the pH change.
*To find this journal article, go to my.central.edu. Under the QuickLinks tab, click on Library. Click on Databases A-to-Z. Click on ACS Journals. Search for the article.

Experiment 11: Buffer Action

Lab Created by Calvin Bill and Seth Signs
Inspired by Lee D. Hansen and Francis R. Nordmeyer, *Exploratory Chemistry Experiments 1*, Kendell/Hunt Publishing Company, Dubuque, Iowa, 1994.

Background:

     A buffer is any solution that resists a change in pH because it contains a weak acid and its conjugate base. When an acid or base within the range of a given buffer is added to the buffer, equilibrium reactions occur and the pH remains relatively unchanged. Added base is neutralized by the weak acid while added acid is neutralized by the conjugate base.

     Buffers are found many places. The shampoo we use, our blood, the fermentation of alcoholic beverages and many other commercial and common processes utilize buffers in order to function. In this experiment, the effects of the addition of strong acids and bases to buffered solutions will be observed. A natural pH indicator derived from isopropyl alcohol and red cabbage will be used to determine pH changes.

Introduction:

Write the Henderson Hasselbalch for buffers of CH_3COONa and CH_3COOH, NH_4Cl and NH_3 and $NaHCO_3$ and H_2CO_3

Other than the buffers used in this lab, describe a one found in everyday life.

Hazards:

Look up and record the hazards of all materials used in this lab.

Procedure:

1. Obtain a test tube rack and small test tubes.
2. Prepare 50 mL of your natural pH indicator using 2 or 3 leaves (6 grams) of red cabbage. Using gloves, tear the cabbage into small pieces about the size of a penny. Place the shredded cabbage into a small beaker and add rubbing alcohol (isopropyl alcohol) until the cabbage is completely submerged. Stir and probe the cabbage solution with a glass stir rod until the rubbing alcohol has started to become dark pink then set the beaker to the side and allow the cabbage to soak further.
3. Fill test tubes 1-3 with 4 mL of DI water.
4. Collect a series of new test tubes. You will use these to observe the pH of the indicator at various pHvalues. With the series of buffer solutions provided, determine the color of the cabbage indicator at pH values from 3-11 by mixing 4

mL of buffer and 4 mL of cabbage indicator in a series of test tubes. Label carefully and set aside.

5. In three test tubes, fill the bottom of each tube with 300 mg of sodium acetate crystals (CH$_3$COONa) before adding 4 mL of vinegar (acetic acid/CH$_3$COOH). This is a buffer.)
6. In three more test tubes, fill the bottom of each tube with 300 mg of ammonium chloride crystals (NH$_4$Cl) before adding 4 mL of household ammonia (NH$_3$). This is a buffer.
7. Finally, in three other test tubes, fill the bottom of each tube with 300 mg of sodium bicarbonate (NaHCO$_3$) before adding 4 ml of DI water. This is also a buffer.
8. Using a vortex machine or by covering the tops of test tubes with parafilm, mix the contents of each test tube until no solid particles remain.
9. Using a dropper pipet, add 3-4 mL (about one full pipet) of the cabbage pH indicator to each test tube and mix the contents again to obtain homogeneity.
10. Observe the color of each test tube and use your knowledge of the natural pH indicator to determine their approximate pH.
11. Add one drop of 1 M HCl (provided) to a test tube containing water/indicator and to one of each set of test tubes from 5,6, and 7.
12. Add one drop of 1 M NaOH (provided) to the water/indicator test tube and to one from each set of buffers (tubes from 5,6, and 7)
13. Observe any and all color changes that occur with the addition of a strong base or acid and comment on the relevance of a color change or lack of color change.
14. After recording your results, dispose of the fluid in each test tube and place the used test tubes in the glass disposal box or rinse to recyle.

Hint: it might be helpful to fill out the following table:

Contents	Color	pH
Water + indicator		
Water+ indicator + HCl		
Water+ Indicator + NaOH		
Indicator + buffer		3
Indicator + buffer		4
Indicator + buffer		5
Indicator + buffer		6
Indicator + buffer		7
Indicator + buffer		8
Indicator + buffer		9
Indicator + buffer		10
Indicator + buffer		11
Indicator + acetate/acetic acid		
Indicator + acetate/acetic acid + HCl		
Indicator + acetate/acetic		

acid +NaOH		
Indicator + ammonia/ammonium		
Indicator + ammonia/ammonium +HCl		
Indicator + ammonia/ammonium +NaOH		
Bicarbonate solution		
Bicarbonate solution +HCl		
Bicarbonate solution + NaOH		

Questions to Discuss in your Results:

1. What happened with the addition of a strong acid or base to the sodium acetate, ammonium chloride and sodium bicarbonate solutions?
2. Why does the initial pH of each buffered solution vary? Consider the pK_a value of 4.75 for CH_3COOH, 9.25 for NH_4^+, and 10.3 for sodium bicarbonate.
3. The pH of the water in test tube 2 should be slightly acidic (below 7.0 pH). Explain this using the fact that the natural indicator does not affect pH and there is a common acidic gas present in the air.
4. How does sodium bicarbonate form a buffer?

Experiment 12: Determination of Sodium by Flame Emission

Adapted from: Analytical Chemistry Sixth Edition by Gary D. Christian

Background:
The intensity of sodium emission at 589.0 nm in a lean, blue flame will be compared with emission of sodium standards. This emission is caused when an excited 3p electron returns to the 3s ground state. A liquid such as Gatorade will be used as the sample. Tap water will also be used as a sample. Students will construct a calibration curve and make sure that the samples tested fall within the calibration curve.

Review: Physical methods came about in the 20th century. They are not based on chemical changes of the analyte.

Atoms emit line spectra. The energy frequency or wavelength of emission will give qualitative information. How much energy is emitted gives quantitate information about the concentration of the analyte.

Reagents:
1. *Stock standard NaCl solution (100 ppm Na).*

Dry about 1 g NaCl at 120°C for 1 hour and cool for 30 minutes. Weigh about 0.254 g NaCl and dilute to 1 L. If you use a different amount, calculate your ppm Na based on the actual amount used. Sodium is the 4th most common element here on Earth so be sure to rinse all your glassware with DI water.

Working standard solutions.
   1. A direct-intensity instrument will be used. Be sure to autozero!

Prepare standards of 0, 1, 5, 10, 15, and 20 ppm Na by diluting 0, 1, 5, 10, 15, and 20 mL of the stock NaCl solution to 100 mL.

b. *Gatorade solution.* Prepare 2, 3, and 4 mL Gatorade in 100 mL volumetric flasks.

c. *Tap water.* Run a sample of tap water directly without dilution. Then, dilute with DI water as needed so that the results fall on the calibration curve. Make sure to have three trials. Adjust aliquot size to fit on calibration curve

Things to Do Before the Experiment:
Dry the NaCl at 120°C for 1 hour and cool in a desiccator. This may have been done ahead of time by the instructor.

Review:
Rinse volumetric glassware with DI water before use.
Fill until meniscus is to line.
Be sure to mix solutions well. Invert 3-10 times
Rinse pipet with solution to be transferred. Allow TC pipets to drain for three seconds while touching to side of beaker or flask.

Procedure:

1. Dilute your samples and standards to volume with water. Aspirate each standard and the unknown and record the emission intensity readings at 589.0 nm. Record instrument type and parameters in your lab notebook. Comment on why the flame should be lean and blue and note the gases used to create the flame.

Using a spreadsheet, plot the emission readings for the standards against concentration and determine the concentration in the unknown solution from the calibration curve plotting concentration vs intensity of emission. From this, calculate the ppm of sodium in the Gatorade and the water. *Don't forget to account for dilution.* Compare to the amount listed on the Gatorade bottle and calculate your absolute error. For the water standard, comment on if this is an expected value for tap water.

Experiment 13: Determination of Dyes in Kool-Aid Using Visible Spectroscopy

Created by: Hailey Benson and Anna Pierce with Dr. Catherine Haustein and with assistance from the Monticello College Foundation.

Background:

Organic compounds that absorb in the visible region of the electromagnetic spectrum have structures that contain non-bonding and pi electrons. In this experiment, the content of artificial colors in consumer products will be determined using visible spectroscopy.

Review:

Beer's law is A = abc

Red 40 Structure 496.4 g/mole

Structures from: Wikipedia

Blue 1 (Brilliant Blue) Structure 792.85 g/mol.

Procedure:
1. *Red 40 Standard.*
    1. The professor will have prepared the standard for you. (250-350 mg Allura Red AC in 1.0 L DI water)
    2. Take aliquots of the standard in increments of .5, 1, 2 5, and 10 mL aliquots in 100 mL volumetric flasks as necessary for the calibration curve.
    3. Test each aliquot in the UV-Vis Spectrometer. Scan the visible region between 400-700 nm. Record absorbance at 509 nm, the "maximum wavelength" where the absorbance is the highest.
    4. Create calibration curve with ppm on x axis and absorbance on y axis. Show trendline and equation. Print curve.
    5. Find unknown ppm of Kool-Aid samples tested (from part 2 of the lab).
2. *Blue 1 Standard.*
    1. Using a volumetric pipet, obtain 2 ml of the Brilliant Blue standard (Blue No. 1), as prepared by your professor (100mg/100ml DI water). Place in 100 ml volumetric flask and dilute with DI water. Invert three times.
    2. Take aliquots of 5, 15, 25, and 35 ml from the solution in step 1 and place in 4 separate 100 ml volumetric flasks. (Note: we are making serial dilutions.) Dilute with DI water. Invert.
    3. Scan the visible region and determine the maximum absorbance wavelength. Record absorbances at 630 nm.
    4. Make calibration curve with concentration in ppm on the x axis and absorbance on the y axis. Print curve.
3. *Kool-Aid samples (Strawberry, Pink Lemonade, Grape, Blue Raspberry, and Black Cherry).*

Make the same concentration in 100 mL as is in 1 pitcher (2 quarts) of Kool-Aid. Note the weight in grams on the packet. (Use conversions. Hint: 1 pint = 473 mL) Check calculation with your professor before continuing.

   1. Use a glass weigh boat to measure amount of Kool-Aid found in step 1. Add to a 100 mL volumetric flask. Dilute with DI water. Note the color of solution.

With the strawberry Kool-Aid, take one aliquot of 10 mL to dilute in a 100 mL volumetric flask. (Measure directly for grape, pink lemonade, and black cherry and dilute if needed.) Alternatively, prepare a pitcher of Kool-aid, diluting a package and taking a 25 mL aliquot diluted to 100 mL as a sample.

Take the Visible spectrum. Record absorbance at 509 and 630 nm. (remember to do more than one sample)

   2. **Conclusion:**
1. Compare grape and black cherry, specifically looking at the ingredient list (type and amount of dye(s)). What dyes are in each and what is the concentration?
2. Compare strawberry and pink lemonade (color and ppm found). Which color has the most dye?
3. Calculate the molar absorptivity (ε) of each dye.
4. Don't forget to include hazards for this lab.

Experiment 14: Two Colorful Plant Molecules--Determination of Beta-carotene Using Visible Spectroscopy/ Identification of Betatin by Fluorescence

Created by: Hailey Benson and Anna Pierce with Dr. Catherine Haustein and with assistance from the Monticello College Foundation.

Pre-Lab/Background:
 Discuss at least three different foods that beta-carotene is found in and three different medical uses of beta-carotene. Discuss the structure of beta-carotene and what structural components make it colored.

 In this lab, a standard of beta-carotene will be used to determine the amount of beta-carotene found in a vitamin supplement. Beta carotene's structure will then be compared with another substance that gives food a color. Finally, fluorescence will be used to confirm the presence of betatin in beets.

Review:
Know the hazards of the chemical you are working with;
refer to the chemical's Material Safety Data Sheet.
When using a fume hood never allow your head to enter the plane of the hood opening.
Use appropriate eye protection.
Wear gloves.

Procedure:
1. *Beta-carotene standard.*
    1. Beta-carotene is stored in the freezer.
    2. Measure 0.025 g beta-carotene on the analytical balance using a glass weigh boat and add to a 100 mL volumetric flask. Dilute to the line with **acetone**. Invert for up to 5 minutes making sure all of the beta-carotene is dissolved. If necessary, use the sonicator with heat for 5 minutes.
    3. Using a volumetric pipet, obtain 0.5 ml of the beta-carotene standard. Place in 100 ml volumetric flask and dilute with **acetone**. Invert. Repeat with 1 mL, 3 mL, and 5 mL volumetric pipets.
    4. From the 3 mL aliquot of the standard (in part c), use a volumetric pipet to take 40 mL, place in a 100 mL volumetric flask, and dilute with **acetone**. Invert.
    5. From the 5 mL aliquot of the standard (in part c), use a volumetric pipet to take 40 mL, place in a 100 mL volumetric flask, and dilute with **acetone**. Invert.
    6. Go to the UV-Vis. Use a **quartz** cuvette (these are expensive, so be careful!). Test the solutions from the volumetric flasks from steps **c-e** (you should test 6 solutions). Use a transfer pipet to transfer the liquid from the volumetric flask into the cuvette. Test all 6 solutions at 430, 455, and 485 nm. Record all absorbances.
    7. Make calibration curve with concentration (ppm) on the x-axis and absorbance (AU) on the y-axis. Make a separate calibration curve for 430,

455, and 485 nm. Show the trendline equation and r^2 value on each curve. Print curves.
2. *Beta-carotene vitamin supplement.*
    1. Take the 100 mL volumetric flask, a glass weigh boat, 1 beta-carotene pill, and a needle into the balance room.
    2. Weigh whole pill on a tared weigh boat. Record mass.
    3. Puncture the pill using the needle. Squeeze contents (a few drops) into 100 mL volumetric flask. Be careful not to get it on your hands! Rinse with acetone to remove all content.
    4. Weigh empty pill casing on the same tared weigh boat. Record mass.
    5. Add roughly 20 mL of **acetone** to the 100 mL volumetric flask. Shake vigorously. Repeat with another 20 mL, shake vigorously. Dilute to line with **acetone**. Invert.
    6. After sample is thoroughly mixed and dissolved, take 1 mL of the sample using a volumetric pipet and add to a 50 mL volumetric flask. Fill to line with **acetone**. Invert.
    7. Go to the UV-Vis. Use transfer pipet to put the solution into a **quartz** cuvette. Wipe cuvette with lens paper. Place in instrument, press sample, record absorbances at 430, 455, and 485 nm. Print spectra.
3. *Beets.*
    1. In this portion of the lab you will be testing beet juice from a can of sliced beets and red beet extract (betanin) from Sigma-Aldrich. Make a prediction about whether or not the beet juice from a can and the red beet extract will contain beta-carotene visible spectroscopy peaks.
    2. For the beet juice, take one transfer pipet full of juice and transfer it to a 100 mL beaker. Dilute with about 80 mL of DI water.
    3. For the red beet extract, measure 0.1 g and add to a 50 mL volumetric flask. Dilute with DI water.
    4. Go to the UV-Vis. Use transfer pipet to put the solution into a cuvette. Wipe cuvette with a lenswipe. Place in instrument, press sample, record peaks and absorbances at the λ_{max} peak(s) in the spectra. Print spectra.
    5. Research what chemical gives color to beets. Obtain this chemical (Betanin) and prepare a solution of 0.1 g in 100 mL of water. Does the spectrum match that of the beets?
4. *Beets and the Fluorimeter.*
    1. Ask instructor before continuing to this portion of the lab.
    2. Prescan—with a solution of betatin.
        1. Set excitation to 482 nm, emission to 545 nm (500-600 nm), and intensity to 168.

c. Scan red beet extract (betatin) with the following settings: excitation 482 nm and emission 544 nm (500-600 nm). Print spectra.
d. Scan beet juice with the following settings: excitation 482 nm, emission 560 nm (500-600 nm. Print spectra.
e. Can you confirm that betatin is in beets?
f. How do the structures of beta carotene and betatin differ?

Calculations:
1. Find the amount of beta-carotene in the vitamin supplement in milligrams and International Units (IU) using the standard calibration curve that has the best r^2 value. Hint: 1 IU = 0.6 μg.
2. Calculate the percent error comparing your results from calculation 1 and the amount in each pill.
3. Calculate the molar absorptivity of beta-carotene based on your measurements and compare it with the known value.

Report:
1. Discuss your findings in step 1 and 2 of the procedure and calculations.
2. Discuss your prediction from step 3 and what you actually found with the beet samples. What can you conclude about the peak(s) in the beet juice and the red beet extract?
3. What can you conclude about the data from the fluorometer?
4. Discuss your accuracy for the beta-carotene measurements.

Experiment 15: Determination of Henna in a Hair Product Using Visible Spectroscopy

Created by: Hailey Benson and Anna Pierce with Dr. Catherine Haustein and with assistance from the Monticello College Foundation.

2-Hydroxy-1,4-napthoquinone (henna)

Pre-Lab/Background:
Research henna and give a brief description of its uses.
In this lab, a standard of henna will be used to determine the milligrams of henna found in hair dye.

Remember to look up hazards and to wear gloves and safety eyewear.

Procedure:
1. *Henna standard.*
    1. Measure 0.01 g 2-Hydroxy-1,4-napthoquinone (henna standard) on the analytical balance using a glass weigh boat and add to a 100 mL volumetric flask. Dilute to the line with DI water. Invert for up to 5 minutes. Sonicate for 15-20 minutes and heat for 30-60 minutes in the sonicator.
    2. Using a volumetric pipet, obtain 10 ml of the henna standard. Place in 100 ml volumetric flask and dilute with DI water. Invert. Repeat with 30 and 50 mL of the henna standard.
    3. From the 50 mL aliquot of the standard (in part b), use a volumetric pipet to take 30 mL, place in a 100 mL volumetric flask, and dilute with DI water. Invert. Repeat with 50 mL.
    4. Go to the UV-Vis. Test the solutions from the volumetric flasks from steps **a-c** (you should test 6 solutions). Use a transfer pipet to transfer the liquid from the volumetric flask into the cuvette. Test all 6 solutions at 338 nm. Record all absorbances.
    5. Make calibration curve with concentration (ppm) on the x-axis and absorbance (AU) on the y-axis. Print curve.

2. *Henna hair dye sample.*
a. Take the 100 mL volumetric flask, a glass weigh boat, and the sample into the balance room.
b. Measure 0.1 g henna hair dye sample on a tared weigh boat. Record mass.
c. Add to 100 mL volumetric flask and dilute with DI water.
d. Sonicate with heat for at least 30 minutes.
e. After sample is thoroughly mixed, go to the UV-Vis. Use transfer pipet to put the solution into a cuvette. Wipe cuvette with a lens wipe. Place in instrument, press sample, record absorbance at 338 nm. Print spectra.

Calculations:

1. Find the amount of henna in milligrams using the standard calibration curve. Calculate the w/w% of henna in the hair dye.

Experiment 16: Determination of Riboflavin (Vitamin B_2) by Fluorescence

Adapted from: Analytical Chemistry Sixth Edition by Gary D. Christian

Pre-lab/Background:
Fluorescence is a sensitive and selective method based on absorbance and emission of uv-vis radiation. Riboflavin is strongly fluorescent in 5% acetic acid solution. White vinegar contains 5% acetic acid and may be used for all dilutions. The riboflavin in an unknown (a vitamin pill) will be determined by comparison of fluorescence intensity to standards. Riboflavin has two excitation wavelengths and we will use the excitation wavelength at 351 nm and the emission wavelength of around 530 nm. (We will determine this maximum emission wavelength as the lab proceeds.)

Look up and record the importance of and the structure of riboflavin in your lab notebook. Comment on the connection between molecular structure and fluorescence.

Hazards:
Look up and record the hazards for riboflavin and acetic acid.

Reagents:
1. *Riboflavin standards.*
    1. Prepare an approximately 90-ppm riboflavin stock solution by accurately weighing about 40 -45 mg riboflavin (to 3 significant figures), transferring to a 500-mL volumetric flask, and diluting to volume with 5% (vol/vol) acetic acid. **White vinegar can be used for the acetic acid.**
b.   Calculate the concentration of riboflavin based on your weight and volume.
c.   This solution should be stored in dark bottle if it is to be used beyond one day.
d.   Dilute an aliquot of this 1:10 to obtain an approximately 10-ppm standard solution.
e.   Dilute aliquots of this with 5% acetic acid to prepare five standards of around 0 (blank), to 8.0 ppm riboflavin. (Calculate the concentration to 3 significant figures.)
f.   Note: 5 cm slits work best for these concentrations.

Procedure:
1. Pre-scan using a standard to confirm the best excitation wavelength and best emission wavelength and to set the instrument slit width.
2. Read the fluorescence of the other standards and prepare a calibration curve by plotting your ppm vs fluorescence intensity (I).
3. Weigh a riboflavin pill on an analytical balance, dissolve approx. a third to a half (weighing carefully) in a 500 mL volumetric flask and dilute to volume with 5% acetic acid. Filter or sonicate if cloudy or let settle and take the top portion.
4. Dilute to obtain a fluorescence intensity that is on your calibration curve.
5. Record fluorescence and calculate the mg riboflavin in the unknown using your curve. Don't forget to take into account the dilutions and fraction of pill sampled.
6. Compare results to that stated on bottle of riboflavin. Calculate the absolute deviation, relative deviation, and standard deviation.

Experiment 17: Determination of Nitrates NO_3^-: UV screening and ion selective electrode

Background:
In this lab we will compare two methods of analysis for the same sample of water. A sample of natural water or tap water will be collected. Nitrate ions are always soluble which means that they can make their way into our water system easily. Nitrates can be determined by their ultra-violet absorbency compared to a calibration curve. Organic materials can interfere so the nitrate content must be determined by using this equation.

$$A_{nitrate} = A_{220\,nm} - A_{275\,nm}$$

Nitrate selective electrode are sensitive and selective for the nitrate ion. The response is based on this equation:

$E = k + 0.0592/n\ p[\text{nitrate activity}]$
n=charge on ion (-1 for nitrate)
X=ion concentration (NO_3^-)

Note: Make sure to use the DI water for dilutions and the tap water or other water for samples. Add the formula for the nitrate ion to your lab notebook.

Hazards:
1. Be sure to use eye and skin protection for this lab. Do not ingest the solutions.

Review:
Rinse with DI water before use.
Fill until meniscus is to line.
Be sure to mix solutions well. Invert 3-10 times
Rinse pipet with solution to be transferred. Allow TC pipets to drain for three seconds while touching to side of beaker or flask.
Rinse electrodes and blot dry before inserting in new solution.

Provided Solutions:
1. Standard nitrate solution.
    1. * This solution will contain 0.8 g KNO_3/ L but be sure to note the exact amount and the volume and calculate the ppm NO_3 and ppm N yourself.
2. 2 M $(NH_4)_2SO_4$ ionic strength adjuster

Procedure:
1. *UV screening.*

1. Using the standard solution provided*, make 3-5 dilutions covering the 0.1 –50 nitrate ppm range or 1-10 ppm nitrate as nitrogen range. Use pipets and volumetric flasks to make the dilution with DI water. Create a calibration curve by measuring absorbance between 200-300 nm in a quartz 1 cm cell.
2. Measure the absorbance of your water sample (tap or natural water) between 200-300 nm. Use the absorbance to determine the nitrate concentration. You will want to make three trials of your water sample. Use the calibration curve to calculate the concentration of nitrate in your water (reported as parts per million nitrogen as nitrate).

2. *Nitrate electrode.*
    1. In this part of the lab we will use an ion selective electrode (ISE). Electrodes respond to ion activity so an ionic strength adjuster (ISA) consisting of 2M ammonium sulfate will be prepared to add to each solution.
    2. Using the stock solution provided, prepare dilutions as before with DI water.
    3. Use volumetric pipets to measure 50 mL of each dilution and 0.5 mL ISA. Mix in a 100 mL beaker.
    4. Immerse the nitrate combination electrode into the solution.
    5. Set the meter to mV display.
    6. Record the mV reading once it has stabilized. Remove electrode, rinse with DI water and blot dry with a Kimwipe. To minimize carryover of sample, measure the most dilute solution first.
    7. Use volumetric pipets to measure 50 mL of the water sample and 0.5 mL ISA as before. Mix in a 100 mL beaker.
    8. To expand your calibration curve, repeat with two additional more concentrated samples, including your undiluted stock solution and one solution made from a 20/100 dilution.
    9. For the calibration curve make a plot of mV vs **log** of nitrate concentration of the standards.

Report:
1. Your lab report introduction and conclusion should include details about nitrates in the environment and may include quotes. You should include a reflection about the significance of nitrate, particularly as a pollutant, and consider why there is an EPA limit for drinking water. Nitrate content of water is usually reported as parts per million nitrogen as nitrate.
2. Compare the sensitivity and selectivity of the two methods.

Experiment 18: Gas Chromatographic Determination of Alcohol in Orange Extract

Created with assistance from the Monticello College Foundation.

Background: Gas Chromatography is used for the determination of volatile compounds in complex mixtures. Like all chromatography, it relies on a stationary phase and a mobile phase. Orange extract is a tincture of orange oil. Orange oil contains primarily d-limonene. Before the lab, look up the structure, hazards, and boiling points of ethanol and d-limonene. Which one should come off first from the non-polar column? Also research and discuss the Thermal Conductivity Detector and its response.

Materials: This lab uses an Agilent GC with a TCD and was developed using Watkins orange extract.

Procedure:

Use the program ORNG with the following settings

Stationary phase: Equity 1 non-polar column
Mobile phase: ultrapure helium
Detector: TCD

Directly inject 1.25 µL Watkins Orange Extract. Record retention times, peak area, and percentage of each peak.

Inject 1.25 µL pure ethanol to identify the alcohol peak

Program ORNG has the following parameters.

Front inlet temperature	150°C
Front Detector temperature	250 °C
Front Detector make up flow	5.3
Flow rate	2 mL/minute

Temperature program
45 °C for 2 minutes
Ramp at 8 degrees per minute up to 114°C

In your discussion, speculate as to the identity of the non-alcohol other peaks in the GC.
Optional exercises:
Verify the identity of a second peak
Change flow rate and compare retention times
Calculate resolution of adjacent peaks
Repeat using an oil instead of an extract. Ramp to 160 °C
Develop a program to get the late eluding peak to come off in under seven minutes.
Create a calibration curve based on volume/volume%

Experiment 19: Gas Chromatographic determination of Methyl salicylate in Oil of Wintergreen and/or Limonene in lemon oil by Gas Chromatography

Developed by Phillip Garcia

Background: Wintergreen oil is derived from two species of wintergreen plants, *Gaultheria procumbens* and *Gaultheria fragrantissima*. The active ingredient, methyl salicylate, is closely related to Aspirin and also contains anti-inflammatory properties. Limonene is the major component of oil in orange peels. Often referred to as d-limonene, it is used as a fragrance ingredient for cosmetic products. Before the lab, look up the structure, hazards, and boiling points of ethanol and methyl salicylate/d-limonene. Which one should come off first from the polar column?

Materials: This lab uses an Agilent GC with TCD (GC2) and a pure Lemon Extract/ wintergreen oil.

Procedure:

Use the program WINTERG with the following settings

Stationary phase HP-5 (crosslinked 5% PH ME Siloxane) polar column
Mobile phase ultrapure helium
Detector TCD

Directly inject 1. 25 µL Extract. Record retention times, peak area, and percentage of each peak.

Inject 1.25 µL pure ethanol to identify the alcohol peak
Inject either methyl salicylate or limonene to identify the fragrance component.

Program WINTERG has the following parameters.

Front inlet temperature	225 °C
Front Detector temperature	250 °C
Front Detector make up flow	5.7
Flow rate	1.3 mL/ minute
Temperature program	150 °C Ramp at 10 °C degrees per minute up to 200 °C

For further investigation:

Extract the lemon or wintergreen scent from a natural source such as lemon peel or a wintergreen plant. Can you detect the component? What solvents work best for the extraction?

Experiment 20: Determination of L-dopa in *Mucuna pruriens* (Velvet beans) by HPLC

Lab Created by Hailey Benson, Nathaniel Holte, Samantha Wills, and Aleesha Godwin with Cathy Haustein

Background:
L-dopa is a naturally occurring drug used in the treatment of Parkinson's disease, a neurodegenerative disease that disables movement. Parkinson's is connected with abnormalities in neurotransmitter metabolism caused by low levels of dopamine in the brain. Dopamine is not readily absorbed across the blood-brain barrier. L-dopa, the precursor to dopamine, can cross this barrier and undergo decarboxylation to form dopamine. L-dopa is a water-soluble weak acid with a pKa of 8.72. In this lab, the pH is adjusted to permit optimum separation of dopamine from L-dopa, which is present in greater amount.
1. Hazards
    1. L-dopa is a health hazard of 1, however, it can cause toxicity so handle with care.
    2. The citric acid buffer may contain a small amount of preservative—sodium azide—that has a health hazard of 3 and a reactivity of 3. It also contains citric acid (health hazard of 2) and sodium phosphate dibasic dihydrate (health hazard of 1.)

Provided solution:
1. *Citric acid buffer.*
    1. Combine 33 g citric acid monohydrate and 2 g sodium phosphate dibasic dihydrate in 800 mL of water. The pH should be around 2.4. If not, add sodium phosphate to increase or citric acid to decrease. If this solution is to be stored for more than a few days, add 18 drops of sodium azide (Health hazard 3! So wear gloves and use hood) and dilute to 1 L.

Standard and dilutions:
2. *L-dopa standard.*
    1. Accurately measure 0-.20 to 0.25 g of L-dopa using the analytical balance and glass weighing funnel. Transfer to a 250 mL volumetric flask by rinsing the funnel with water. Add 20 mL of the pH 2.3-2.4 citric acid buffer before diluting to the mark. Based on your actual weight, calculate the ppm of this solution.
    2. Sonicate or heat mildly and stir.

Procedure:
1. *Dilutions.*
    1. Make four dilutions of the L-dopa to cover the concentration range of 500-100 ppm. Use 100 mL volumetric flasks.
    2. Filter each solution with a syringe filter and a 0.5 μm disposable filter.
        1. You will only need to filter a syringe full!
c.   After setting up the HPLC, inject 20 μm of standard.
HPLC.
.   Program name: dopamine

                1. Isocratic run of pH 2.4 citric acid buffer for 12 minutes, 1mL/min
a. Mobile phase: 2.3 -2.4 pH citric acid buffer
b. Stationary phase: phenomenex Luna 5u C-18 250 x 4.60 mm 5 microns
c. UV detection at 280 nm
d. Temperature: 30 °C

Velvet beans. Weigh approximately 3 grams of velvet beans on an analytical balance. Soak in about 300 mL of water until next lab period.

---**Day 2 of Lab**--a.
a. Filter bean water into a 500 mL volumetric flask using filter paper.
b. Add about 20 mL of pH 2.4 buffer and dilute to the mark.
    c. Filter with syringe filter as before and run under previous conditions. Repeat
    d. Combine class data to construct calibration curve and determine the w/w % L-dopa in the beans. For the calibration curve plot area (y axis) vs ppm (x axis)
    e. *Discussion and conclusions.*

What is the structure of L-dopa?
Are your results in alignment with those previously published?
How might changing the pH of the mobile phase affect the retention time of L-dopa?

APENDIUM ON L-DOPA METHOD
Table 1. Previous results

	Days soaked	L-dopa %	Stdev
Individual sp 15	2	3.8%	0.1
Chem 241 f 14	2	3.7%	2.5
Chem 442 sp 15	2	3.4%	0.03
Chem 241 sp 15	3	5.5%	0.3
Chem 241 sp 16	2	3.41%	0.5
Chem 442 sp17	2	3.87%	0.03
Chem 442 sp 21	2	4.02%	0.2

Table 2. Most to least effective method of extraction

Method	% L-dopa
Three day whole beans at room temperature	4.69%
Three day ground beans at room temperature	4.09%
Two day whole beans at room temperature	3.76%
Two day boiled beans	3.48%
Two day whole beans refrigerated temperature	2.86%
Two day ground beans at room temperature	2.11%
Two day whole beans in ethanol	0.88%

Experiment 21: Determination of the slope of a pH electrode

Review: Potentiometric determination of pH utilizes a silver/silver chloride reference electrode and a pH electrode with a glass membrane sensitive to hydrogen ion activity. Ideally, a pH electrode will measure 0 mV at pH 7.0 and have a slope of 59.2 at 25 °C.

$$E = k + 0.0592/n \text{ pH}$$

n=charge on ion (+1 hydrogen ion)

Procedure:

Locate two pH/ reference electrode combination probes and one pH meter.

Create a series of buffers from pH 2-11 using buffer pills or premade buffers.

Prepare a 1 M solution of ionic strength buffer made from KCl. Check hazards for KCl!

For each solution tested, mix 50 mL buffer and ½ mL KCl.

Measure the temperature of each solution. Measure the mV using one pH electrode. Create a plot of mV vs pH. Repeat with the second electrode. Evaluate each electrode.

Experiment 22: Determination of Quinine by Fluorescence

Adapted from: Donald T. Sawyer, William R. Heineman, Janice M. Beebe, <u>Chemistry Experiments for Instrumental Methods</u>, Experiment 10-1, 271-273, 1984.
and http://webpages.acs.ttu.edu/dpappas/index.html

Quinine structure

Purpose:
Quinine is a strongly fluorescent compound in dilute acid with two excitation wavelengths (250 and 350 nm) and emission at 450 nm. We will determine the amount of quinine in tonic water. Heavy atom quenching's affect on the fluorescence of quinine will be studied as well as the interpretation of the emission spectra.

Theory:
For low concentrations, the fluorescence emission intensity is directly proportional to the concentration as well as to the intensity of the excitation following this equation
$F = P_o(2.3\ \varepsilon bc)Qk$ holds generally for concentrations up to a few micrograms per milliliter.
Where
    F = intensity of fluorescence
    ε = molar absorptivity
    b = path length
    c = concentration
    P_o = power of incident radiation
    Q = quantum efficiency = photons emitted/photons absorbed
    k = photons measured/photons emitted

Reduction in intensity of fluorescence is called quenching. It can be due to factors such as promotion of inner system crossing (to the excited triplet state) which is unstable in solution due to collisions with solvent molecules. This is known as collisional quenching. *Concentration quenching* (a decrease in the fluorescence as the concentration is increased), and *chemical quenching* due to changes in structure caused by factors such as pH. Concentration quenching results from excessive absorption of fluorescent radiation.

<u>Excitation and Emission Spectra</u>
The fluorescence emission spectrum of an organic molecule is similar to its excitation (absorption) spectrum.

The types of transitions associated with fluorescence include π-to-π* and π*-to-π and n to π* and π* to n.

Under usual conditions and at a concentration of about 2 µg/mL quinine, the two excitation peaks (250 and 350 nm) and one emission peak (450 nm) will be seen with quinine.

Procedures

A. <u>Preparation of a Calibration Curve:</u> Determination of Quinine in Unknowns and Limit of Detection
- Prepare a quinine stock solution: (100.0 µg/mL quinine) from 120.7 mg quinine sulfate dihydrate or 100.0 mg quinine sulfate, transfer to a 1-L volumetric flask. Dilute to mark with white vinegar.

Prepare a series of quinine standards from the 100-µg/mL stock standard. Make sequential tenfold dilutions with vinegar. This should give concentrations of 10, 1, 0.1, 0.01, 0.001, 0.0001 µg/mL, etc. Collect replicate scans of the blank. Determine the detection limit. Make dilutions until the most dilute solution gives a fluorescence intensity approximately that of the blank (vinegar.)

The emission wavelength is the same (450 nm) for either of the excitation wavelengths (250 and 350 nm). Also measure the fluorescence of a sample of quinine tonic water after the following dilutions: pipet 5.00 mL of tonic water into a 250-mL volumetric flask, dilute to the mark with vinegar; pipet 5.00 mL of this solution into a 25-mL volumetric flask and dilute to volume with vinegar.

<u>Treatment of Data</u>

Plot the relative fluorescence intensity versus the quinine concentration. Discuss any deviation from linearity in the plots. Account for deviations if present and delete these points from your calibration curve. Determine the concentration of quinine in the tonic water. (Remember to calculate and report the concentration of quinine in the original tonic water.)

B. <u>Halide Heavy Atom Quenching of Quinine Fluorescence:</u>

Pipet 2.0 mL of the 10 µL standard quinine sulfate solution into a 25-mL volumetric flask, add 1.0 mL 0.05 *M* KI and fill to the mark with vinegar. Prepare four more solutions with 2.0 mL of 10 µg/mL quinine in each and add 2.0-, 4.0-, 8.0-, and 16.0 mL portions of KI. Measure the fluorescence of the five solutions.

<u>Treatment of data</u>

Plot fluorescence intensity versus **concentration** of the heavy ion. Explain the results.

Experiment 23: Determination of Iron in Wine and Beer or Pomegranate Juice by Atomic Absorption (New lab)

Iron is found in wine at a level of around 30 ppm and in beer at a level of around 8 ppm. In this lab we will determine the iron in wine directly and use standard addition to determine the amount of iron in beer since the presence of copper can interfere at such a low level.

Atomic absorption conditions:

Use the iron hollow cathode lamp set at approximately half the maximum amps.

Measure absorption at the 248.8 nm wavelength. Best results are obtained with a 0.2 nm slit size and a 10 second integration time. And a lean flame.

Prepare a standard solution of 1000 ppm using ferrous ammonium sulfate hexahydrate. Accurately weigh 0.702 g ferrous ammonium sulfate (Mohr's salt) into a 100 mL volumetric and dilute. Prepare a calibration curve using 1,2,3,5, and 10 mL aliquots in 100 mL. Aspirate into a lean blue flame. Prepare a calibration curve. Wine should be able to be measured directly without dilution.

For beer, sonicate or shake to remove the foam from one can or bottle of beer. Pomegranate juice may be used in place of beer. Prepare a calibration curve as follows

Add 20 mL of flat beer to each of 6 100 mL volumetric flasks.

Dilute one flask. This is the zero point.

Prepare a calibration curve using the standard iron as added.
Add 1 mL standard to one volumetric and dilute. Next prepare 2,3,5, and 10 mL additions to each of the remaining flasks. Plot using the amount of standard added (the ppm of the standard as diluted) for your x value. Extend x axis into the negative region. The absolute value (where y=0) of the spot where the line crosses the x axis represents the concentration of iron in the beer. (Stroh's Bohemian and Modelo work well for this lab.) Remember to multiply by 100/20 to account for the dilution.
Pomegranate juice also works instead of beer for this lab but check the label to make sure it includes iron.

Experiment 24: Determination of p-coumaric acid in sage using fluorescence and standard addition (Lab Under Development)

Lab created by Taylor Hull and Hailey Benson with Cathy Haustein

Background:
In the standard addition technique, aliquots of the analyte are added to a fixed amount of sample to create a calibration curve. It is used for a complex sample matrix in which the interferences aren't well known. Sage is a complex matrix. The organic acid p-coumaric acid is a plant antioxidant. It's prevalent in corn at a concentration of 4% by weight. It could have anti-cancer properties.

In standard addition, the signal is proportional to the analyte concentration. The equation to apply is
$[PC]_{samp} = [PC]_{standard} \times V_o/V_d$
where
$[PC]$ = concentration of p-coumaric acid
V_o = original volume and V_d is diluted volume

Procedure:
Prepare a standard of approximately 0.1 g p-coumaric acid (weighed on an analytical balance) in 100 mL ethanol. This has the concentration of 1000 ppm.

Prepare pH 10. 2 carbonate buffer (insert instructions here). 42.00 g sodium carbonate in 1 L distilled water. Add 80 mL 1 M NaOH. Adjust pH with 1 M NaOH.

Weigh 1.0-1.2 g sage on an analytical balance. Transfer to a glass screw top vial. Add 10 mL of distilled water with a pipet. Shake or sonicate for 10 minutes. Filter with a Buchner funnel into a 10 mL volumetric flask and dilute to the mark. Shake.

Standard addition curve:
     Transfer a 2.0 mL aliquot to a 10 mL volumetric flask. Dilute with water. Remove 2.5 mL with a volumetric pipet. Add 1.0 mL carbonate buffer. Measure fluorescence emission from 425-600 nm using an excitation wavelength of 404 nm and 20 nm slits.

Transfer a 2.0 mL aliquot to a 10 mL volumetric flask. Add 1 mL p-coumaric acid standard. (100 ppm). Dilute with water. Remove 2.5 mL with a volumetric pipet. Add 1.0 mL carbonate buffer. Measure fluorescence emission from 425-600 nm using an excitation wavelength of 404 nm and 20 nm slits.

Repeat with additions of 2.0 mL p-coumaric acid and 3.0 mL p-coumaric acid.

Plot mL p-coumaric acid added on the x axis and intensity of emission at on the y axis. Extend the x-axis into the negative region. Mark the place where the trend line intercepts the x axis. The absolute value of this number represents $[PC]_f$ p-coumaric acid in the sage with the units of mL. Convert to % p-coumaric acid.

Experiment 25: Identification of Polymers using Infrared Spectroscopy (Lab Under Development)

Review:
1 polyethyleneterephtalate
2, 4 polyethylene
3 polyvinylchloride
5 polypropylene
6 polystyrene

Collect a sample of each recyclable plastic #1-6 and a sample of nylon. Using FTIR, obtain a spectrum of each sample.

Identify the transition responsible for each IR peak.

Identify several unknown samples by matching the peaks.

Experiment 26: Determination of 4-acetamidophenol in analgesics using Cyclic Voltammetry

Developed by Calvin Bill and Seth Signs and the 2017 Instrumental Analysis class

Based on: Electrochemical Analysis of Acetaminophen in a Common Pain Relief Medication **KMITL Sci. Tech. J. Vol. 5 No. 3 Jul.-Dec. 2005**

Background: 4-acetamidophenol is the formal name for the drug acetaminophen, a commonly used analgesic. This structure can be oxidized in an irreversible reaction. The reaction is pH sensitive and a buffer of around pH 7.0 should be used for reproducible results. It's also important to have a supporting electrolyte. Contact lens solution provides a convenient buffer and electrolyte although pH 7.0 phosphate buffer can also be used. This lab is applied to children's medication but standard addition can be used for more complicated samples.

Stock solution Make 50mL of a 35mM stock solution of acetaminophen in contact lens solution.

Dilutions Using the stock solution create 1.4mM, 3.5mM, 7.0mM, and 10.0mM dilutions of 25 mL in volumetric flasks using contact solution as the buffer

Data collection:

At Central College we have a PAR potentiostat and will use a gold working electrode and platinum counter electrode.

Pour the solutions into a small vial. Make sure that the working and counter electrodes are next to each other for best results. Degas by gently blowing nitrogen over the solution for 30 seconds. Allow at least 10 seconds for equilibrium before scanning.

The cyclic voltammograms are obtained by scanning the potentials from 0.0 to +1.0 V to -0.5 and back to 0.0 V vs. Ag/AgCl at step potential of 0.0005 V and scan rate of 50 mV/s.

Plot both the anodic and cathodic currents on separate calibration curves. Which one shows greater sensitivity?

Dissolve a tablet of children's Tylenol in 50 mL of contact solution. Dilute 10 mL of the solution in 25 mL, again diluting with contact lens solution. Prepare two samples. Obtain the voltammograms.

Use your results and your concentration curve to determine the concentration of acetaminophen and compare to the amount listed on the packaging.

Experiment 27: Determination of Chloride by Ion Selective Electrode

Review:
Potentiometry is an analytical method based on electrical properties of a solution of an analyte when it is made part of an electrochemical cell.
It is characterized by a high degree of sensitivity, selectivity and accuracy.
It require a reference electrode with a fixed voltage, an indicator electrode with a variable voltage, and a potentiometer (a meter which measures voltage.)
The method may be applied to cloudy, turbid, and colored samples.
It has a wide linear range (dynamic range). The response is based on the Nernst equation is due to activity (a) and requires the use of a total ionic strength adjustor. It measures free ions.

Background: The chloride ion selective electrode you will use is a crystalline solid-state electrode. The membrane consists of a solid salt of silver sulfide / silver chloride (Ag_2S / $AgCl$).

The electrode response will be
$E = k + 0.0592/n \, pCl$
n = charge on ion (-1 for chloride)

Procedure:

Be sure to clean glassware thoroughly.

a. ISA—Ionic Strength Adjuster solution (5M $NaNO_3$): Dissolve 42.5 g of sodium nitrate in di water to make 100 mL of solution. Store in a clean glass or plastic container.

b. Chloride calibration stock solution (1,000 mg/L Cl): Dissolve 0.1649 g of sodium - chloride (dried two hours at 110 °C and stored in a desiccator) in "reagent" water and dilute to 100 mL in a volumetric flask. Store in a clean bottle.

Prepare a series of calibration standards covering the 1000-0.5 ppm range.
by diluting the 1,000 mg/L chloride standard. You might need to use serial dilution.

Equilibrate the electrode for at least one hour in a 30.0 mg/L chloride standard before use. Be sure to fill it with electrolyte solution before use.

To measure:
Add 50.0 mL of standard and 1.00 mL of ISA to a 100 mL beaker. Stir completely.

Record the mV reading when stable.

Plot log concentration vs mV.

Unknowns:
Unknowns/samples:
a. Split class into two groups. Create an unknown for the other group using the chloride ion standard.
b. Use contact lens solution as an unknown. Determine if use of EDTA as a masking agent improves results.

Experiment 28: Determination of Fluorescein in AntiFreeze (New Lab)
Developed by Calvin Bill, Seth Signs, and Alexis Oldham

First create stock solutions of .05M and .01M fluorescein in 500 mL of DI water. Then create serial dilutions from the stock solution of 1:25, 1:50, 1:100, and 1:200 using .1M NaOH as the diluent. Analyze the solutions using the fluorimeter with emission values around 513 nm with excitation set to 492. From this data create a calibration curve comparing fluorescein concentration to fluorescence intensity. After creation of the calibration curve dilute two samples of antifreeze 10:100 and 20:100 using NaOH as the diluent to hold pH consistent between the two tests. Analyze these samples using the fluorimeter and use these readings to calculate the percent composition fluorescein in the antifreeze sample and compare your results to generally accepted values which are available online.

Dispose of the antifreeze solutions by pouring them on an absorbent mat and placing in the appropriate waste container.

Experiment 29: Determination of Aspirin by HPLC (New Lab)
Developed by Alexis Oldham and Jamie Heron

Acetylsalicylic acid is the main ingredient in Aspirin. It is an anti-inflammatory used to combat mild pain An HPLC method that was developed by M. Sawyer and V. Kumar was modified to determine the amount of acetylsalicylic acid aspirin tablets (1). A low-dose 81mg Aspirin tablet, a 325mg Aspirin tablet and an 165mg Aspirin tablet from the stockroom were used during this experiment. The reduced dose of acetylsalicylic acid in aspirin tablets can be caused by the hydrolysis of acetylsalicylic acid to acetic acid and salicylic acid. The modifications made to make this a cleaner, greener lab was a change in solvents. Isopropanol and vinegar were used in place of methanol and glacial acetic acid during the experiment.

Parameters:
- Mobile phase- 1/3 Vinegar 2/3 Isopropanol with a pH of 3.52.
- Stationary phase- Phenomenex C-18
- Syringe filtration was performed on the samples before injection into HPLC
- HPLC conditions: Agilent 1100 series pump with internal variable UV wavelength detector (set at 275 nm) and a Rheodyne 7725i injector port with a 20-μL sample loop. Luna C-18 non-polar, reverse-phase column. Column was jacketed to maintain a 30 °C temperature. The flow rate was 1.0 mL/min. The run time was 9 minutes, t_R=6.6.
- All of the vinegar used should be filtered through a Cole-Parmer nylon membrane filter with a filtration rating of .2μm.

Procedure:
- Prepare a standard solution of 1000 ppm acetylsalicylic acid by dissolving 100 mg in 100 mL of mobile phase. Create a calibration curve ranging from 600- 60 ppm via dilution.
- Dissolve aspirin tablet (325mg) in 100mL of the vinegar/isopropanol mobile phase. Ten, 5mL and 4 mL aliquots should be transferred into two 100mL volumetric flasks.
  -

- 1 M. Sawyer and V. Kumar. A Rapid High-Performance Liquid Chromatography Method for the Simultaneous Quantitation of Aspirin, Salicylic Acid, and Caffeine in Effervescent Tablets. Journal of Chromatographic Science. 41:393-397 (2003).
  - www.pubchem.ncbi.nlm.nih.gov

Experiment 30: Determination of Tryptophan by Fluorescence (New Lab)
By D. Wayne Sommerlot Class of 2020

Pre-Lab/Background:
Tryptophan is an essential amino acid that plays a crucial role in many metabolic functions, such as the synthesis of 5-HT serotonin. Humans are unable to produce tryptophan in the body, and therefore must obtain it from their diet. Supplementation of tryptophan can be used as a treatment for sleeping disorders and depression. Tryptophan is found in skeletal muscles. This is one reason you feel happy and drowsy after a large Thanksgiving turkey dinner. Look up other dietary sources of tryptophan and record in lab notebook.

Hazards:
Look up the hazards for tryptophan and acetic acid.

Reagents:
1. Prepare 1000ppm standard in vinegar using provided tryptophan.
2. Using a fluorometer, find the ideal excitation wavelength of tryptophan using prescan. (should get 305 nm excitation and emission between 350-360nm)
3. Take 3-5 dilutions of the stock solution (standard) to create regular intervals from 100ppm to 350ppm.
4. Scan 3 samples of each dilution, average and record in lab notebook.
5. Using the results of your scans, create a calibration curve for tryptophan.

Procedure:
1. Using an analytical balance, weigh out approximately $1/8^{th}$ (0.05g) of a Turkey Sleep pill.
2. Transfer to a 100mL volumetric flask.
3. Dissolve (sonogram as needed) and then dilute to volume with 5% acetic acid (white vinegar).
4. Set the fluorimeter excitation to 305nm, scanning from 320nm to 600nm with a slit of 5nm.
5. Using the high precision cell, zero the fluorimeter with pure solvent (acetic acid).
6. Scan 3 separate samples of the Turkey Sleep solution using the high precision cell, record the peak and intensity readings.
7. Average your readings.
8. Using your calibration curve, find the amount of tryptophan in a single Turkey Sleep pill.
9. Compare results with the Turkey Sleep label and calculate your percent error.
10. Compare and discuss results with classmates, record in lab notebook.

Application:
1. Take a can of mixed nuts (Planters deluxe unsalted used in original study), sort out the different kinds of nuts and label.
2. Using a mortar and pestle, grind 2.5 grams of each type of nut, place in 100ml beaker, cover with 50-70ml of white vinegar.
3. Cover, and let sit for one week.

4. Filter each solution into a 100ml volumetric flask, dilute to volume with white vinegar.
5. Take 3 samples of each solution, measure the fluorescence at the wavelengths determined above, record and find the mean.
6. Discussion: Which nuts contain tryptophan? Which nuts do not contain tryptophan? Do a quick search to verify conclusions.

Experiment 31: Field Lab: Control Chart

In this field lab, you will go to an enclosed space and monitor the carbon dioxide levels.

The outdoor concentration of carbon dioxide can vary from 350-400 parts per million (ppm) or higher in areas with high traffic or industrial activity.

The level of CO_2 indoors depends upon:

- the number of people present
- how long an area has been occupied
- the amount of outdoor fresh air entering the area
- the size of the room or area
- whether combustion by-products are contaminating the indoor air (e.g., idling vehicles near air intakes, leaky furnaces, tobacco smoke)
- the outdoor concentration

Carbon dioxide concentrations indoors can vary from 400 ppm to over 1000 ppm in areas with many occupants present during an extended period of time when outdoor air ventilation is limited.

https://www.health.state.mn.us/communities/environment/air/toxins/co2.html

Evidence suggests that even above 600 ppm, indoor carbon dioxide impairs decision making for some individuals. Above 750 ppm, focused activity may be affected.

Levels above 800 ppm indicated poor air quality and above this, the average individual can become unmotivated and lack clear decision-making capacity.

These levels are well below the danger level for carbon dioxide intoxication. However, for this course, we will assume that any level above 600 ppm is not an acceptable level of air quality and indicates that air in the indoor space is becoming stagnant.

https://newscenter.lbl.gov/2012/10/17/elevated-indoor-carbon-dioxide-impairs-decision-making-performance/

Identify a room where the carbon dioxide will be monitored. Determine how long you wish to monitor.

Humans provide most of the carbon dioxide beyond background levels in a building. Using a simple, portable carbon dioxide meter, take unoccupied measurements in the area to determine a baseline. Get several measurements and determine the baseline standard deviation.

Then note the deviation from the baseline as the room is occupied and used. This deviation can be due to a "determinate error" caused by poor ventilation.

Or your chart, put time on the x axis and deviation from baseline (standard deviation and ppm deviation) on the Y axis. Make note of the acceptable limits of deviation. Determine if and when your enclosed space goes beyond the average deviation and beyond the acceptable level of carbon dioxide.

Turn in a typed report that includes:

An **introduction,** including how the meter operates and detects and measures the carbon dioxide.

A **methods** section explaining how you made your measurements

A **results** section which includes the control chart

A **conclusions** section, indicating the safety and air quality of the space monitored.

APPENDIX:

Common Suffixes Used with Nouns in Chemical Terminology

-ance, -ence: denotes a state, a condition, a quality of being
    Example: valence, conductance, resistance, absorbance

-ant, -ent: denotes a thing acting as an agent
    Example: precipitant, titrant, diluent, dissectant, eluent

-ate: denotes nouns made from participles (verb forms), something resulting from the action of
    the verb stem
    Example: precipitate, distillate, condensate, decantate, eluate, absorbate

-gram: denotes something drawn or written
    Example: polarogram, spectrogram, chromatogram

-graph: denotes an instrument for making records
    Example: polarograph, spectrograph, chromatograph

-meter: denotes an instrument for measuring some specified thing
    Example: thermometer, spectrometer, spectrophotometer, photometer, potentiometer

-or: denotes an agent or doer
    Example: desiccator, monochromator

-tion, -sion: denotes action, a process
    Example: precipitation, distillation, condensation, dissociation, decantation, elution,
        adsorption

-ty, -cy: denotes a quality, a state, a condition
    Example: density, conductivity, resistivity, absorptivity, accuracy

Equipment that makes a measurement as used in our physical methods is an **instrument.** (A machine does work.)

Name		Drawer #		Key ID	
Course	Analytical	Check out Date		Key Returned?	Yes | No
Semester		Check in Date			

Lab Drawer Contents:

Item	Qty	Out	In
Beakers: (Note: blue lettering on beakers is preferred)			
1000 mL	1		
600 mL (1-Kool grip)	4		
800 mL (Kool grip)	1		
400 mL (1-blue lettered)	3		
250 mL (1-blue lettered)	3		
100 mL (1-blue lettered)	3		
10 mL	1		
Plastic Bottle w/Lid:			
1000 mL, white	2		
1000 mL, brown	2		
Graduated Cylinders:			
100 mL	1		
10 mL	1		
Flasks:			
100 mL plastic volumetric with cap	1		
250 mL Erlenmeyer	3		
Watch Glasses:			
~7.5 cm	3		
~12.5 cm	3		
Funnels:			
Long stem, glass	1		
Large Plastic	1		
Other Items:			
Tongs	1		
Dropper pipettes	1		
Forceps	1		
Glass hooks	6		
Glass stir rods	2		
Magnetic stir bar	1		
Spatula	1		
Weighing bottles, glass	3		
Weighing funnel, glass	1		
Litmus paper, red	1		
Litmus paper, blue	1		
Pipet bulb, blue	1		

Beginning of Semester
Check out verified by:

_____

Date:

_____

End of Semester
Check out verified by:

_____

Date:

_____

Verifications should be done by Professor or Lab TA

Additional items may be needed
    Thermometers can be checked out of stockroom **with ID**, when needed

    Desiccator, Sand bath, & power mites are in VSC 280

About the authors:

Catherine Haustein is a professor of chemistry at Central College. She's also the author of the novels *Natural Attraction* (Penner Publishing), *Wolves and Deer* (Rogue Phoenix Press), *Mixed In* and *Lost in Waste* (City Owl Press).

Anna Pierce was a 2016 graduate of Central College majoring in Chemistry, Biology, and Secondary Education. She teaches high school science in Denver.

Hailey Benson was a 2017 graduate of Central College majoring in Biochemistry. She greatly enjoys spending time in the lab as well as being an avid writer. She is employed by the Mayo Clinic.

Nathaniel Holte was a 2016 graduate of Central College majoring in Biochemistry. He is currently a podiatrist.

Zachary M Miller is a 2017 graduate of Central College majoring in Chemistry. He is currently working as an Analytical Chemist.

Calvin Bill and Seth Signs are 2018 graduates of Central College. Calvin is attending Washington University School of Law. Seth is a student at University of Kansas Medical School.

Phillip Garcia was a 2020 graduate of Central College majoring in biochemistry. He currently works for Safe Science.

Wayne Sommerlet graduated in 2020 with a major in environmental studies. He is currently in the Marine Corp officer training.

Thank you to the Margaret A. Cargill Foundation and the Monticello College Foundation for their assistance with this manual.

Please direct any corrections and concerns to hausteinc@central.edu

www.ingramcontent.com/pod-product-compliance
Lightning Source LLC
Chambersburg PA
CBHW080619220526
45466CB00010B/3397